RILEM Recommendations for the Prevention of Damage by Alkali-Aggregate Reactions in New Concrete Structures

RILEM STATE-OF-THE-ART REPORTS
Volume 17

RILEM, The International Union of Laboratories and Experts in Construction Materials, Systems and Structures, founded in 1947, is a non-governmental scientific association whose goal is to contribute to progress in the construction sciences, techniques and industries, essentially by means of the communication it fosters between research and practice. RILEM's focus is on construction materials and their use in building and civil engineering structures, covering all phases of the building process from manufacture to use and recycling of materials. More information on RILEM and its previous publications can be found on www.RILEM.net.

The RILEM State-of-the-Art Reports (STAR) are produced by the Technical Committees. They represent one of the most important outputs that RILEM generates—high level scientific and engineering reports that provide cutting edge knowledge in a given field. The work of the TCs is one of RILEM's key functions.

Members of a TC are experts in their field and give their time freely to share their expertise. As a result, the broader scientific community benefits greatly from RILEM's activities.

RILEM's stated objective is to disseminate this information as widely as possible to the scientific community. RILEM therefore considers the STAR reports of its TCs as of highest importance, and encourages their publication whenever possible.

The information in this and similar reports is mostly pre-normative in the sense that it provides the underlying scientific fundamentals on which standards and codes of practice are based. Without such a solid scientific basis, construction practice will be less than efficient or economical.

It is RILEM's hope that this information will be of wide use to the scientific community.

More information about this series at http://www.springer.com/series/8780

Philip J. Nixon · Ian Sims
Editors

RILEM Recommendations for the Prevention of Damage by Alkali-Aggregate Reactions in New Concrete Structures

State-of-the-Art Report of the RILEM Technical Committee 219-ACS

rilem

Springer

Editors
Philip J. Nixon, Associate
Building Research Establishment
Watford, Hertfordshire
UK

Ian Sims, Director
RSK Environment Ltd
Hemel Hempstead
UK

ISSN 2213-204X ISSN 2213-2031 (electronic)
RILEM State-of-the-Art Reports
ISBN 978-94-017-7251-8 ISBN 978-94-017-7252-5 (eBook)
DOI 10.1007/978-94-017-7252-5

Library of Congress Control Number: 2015945608

Springer Dordrecht Heidelberg New York London

Printed on acid-free paper

Springer Science+Business Media B.V. Dordrecht is part of Springer Science+Business Media (www.springer.com)

RILEM Publications

The following list is presenting the global offer of RILEM Publications, sorted by series. Each publication is available in printed version and/or in online version.

RILEM Proceedings (PRO)

PRO 1: Durability of High Performance Concrete (ISBN: 2-912143-03-9); *Ed. H. Sommer*
PRO 2: Chloride Penetration into Concrete (ISBN: 2-912143-00-04); *Eds. L.-O. Nilsson and J.-P. Ollivier*
PRO 3: Evaluation and Strengthening of Existing Masonry Structures (ISBN: 2-912143-02-0); *Eds. L. Binda and C. Modena*
PRO 4: Concrete: From Material to Structure (ISBN: 2-912143-04-7); *Eds. J.-P. Bournazel and Y. Malier*
PRO 5: The Role of Admixtures in High Performance Concrete (ISBN: 2-912143-05-5); *Eds. J. G. Cabrera and R. Rivera-Villarreal*
PRO 6: High Performance Fiber Reinforced Cement Composites—HPFRCC 3 (ISBN: 2-912143-06-3); *Eds. H. W. Reinhardt and A. E. Naaman*
PRO 7: 1st International RILEM Symposium on Self-Compacting Concrete (ISBN: 2-912143-09-8); *Eds. Å. Skarendahl and Ö. Petersson*
PRO 8: International RILEM Symposium on Timber Engineering (ISBN: 2-912143-10-1); *Ed. L. Boström*
PRO 9: 2nd International RILEM Symposium on Adhesion between Polymers and Concrete ISAP '99 (ISBN: 2-912143-11-X); *Eds. Y. Ohama and M. Puterman*
PRO 10: 3rd International RILEM Symposium on Durability of Building and Construction Sealants (ISBN: 2-912143-13-6); *Eds. A. T. Wolf*
PRO 11: 4th International RILEM Conference on Reflective Cracking in Pavements (ISBN: 2-912143-14-4); *Eds. A. O. Abd El Halim, D. A. Taylor and El H. H. Mohamed*
PRO 12: International RILEM Workshop on Historic Mortars: Characteristics and Tests (ISBN: 2-912143-15-2); *Eds. P. Bartos, C. Groot and J. J. Hughes*

PRO 13: 2nd International RILEM Symposium on Hydration and Setting (ISBN: 2-912143-16-0); *Ed. A. Nonat*
PRO 14: Integrated Life-Cycle Design of Materials and Structures—ILCDES 2000 (ISBN: 951-758-408-3); (ISSN: 0356-9403); *Ed. S. Sarja*
PRO 15: Fifth RILEM Symposium on Fibre-Reinforced Concretes (FRC)—BEFIB'2000 (ISBN: 2-912143-18-7); *Eds. P. Rossi and G. Chanvillard*
PRO 16: Life Prediction and Management of Concrete Structures (ISBN: 2-912143-19-5); *Ed. D. Naus*
PRO 17: Shrinkage of Concrete – Shrinkage 2000 (ISBN: 2-912143-20-9); *Eds. V. Baroghel-Bouny and P.-C. Aïtcin*
PRO 18: Measurement and Interpretation of the On-Site Corrosion Rate (ISBN: 2-912143-21-7); *Eds. C. Andrade, C. Alonso, J. Fullea, J. Polimon and J. Rodriguez*
PRO 19: Testing and Modelling the Chloride Ingress into Concrete (ISBN: 2-912143-22-5); *Eds. C. Andrade and J. Kropp*
PRO 20: 1st International RILEM Workshop on Microbial Impacts on Building Materials (CD 02) (e-ISBN 978-2-35158-013-4); *Ed. M. Ribas Silva*
PRO 21: International RILEM Symposium on Connections between Steel and Concrete (ISBN: 2-912143-25-X); *Ed. R. Eligehausen*
PRO 22: International RILEM Symposium on Joints in Timber Structures (ISBN: 2-912143-28-4); *Eds. S. Aicher and H.-W. Reinhardt*
PRO 23: International RILEM Conference on Early Age Cracking in Cementitious Systems (ISBN: 2-912143-29-2); *Eds. K. Kovler and A. Bentur*
PRO 24: 2nd International RILEM Workshop on Frost Resistance of Concrete (ISBN: 2-912143-30-6); *Eds. M. J. Setzer, R. Auberg and H.-J. Keck*
PRO 25: International RILEM Workshop on Frost Damage in Concrete (ISBN: 2-912143-31-4); *Eds. D. J. Janssen, M. J. Setzer and M. B. Snyder*
PRO 26: International RILEM Workshop on On-Site Control and Evaluation of Masonry Structures (ISBN: 2-912143-34-9); *Eds. L. Binda and R. C. de Vekey*
PRO 27: International RILEM Symposium on Building Joint Sealants (CD03); *Ed. A. T. Wolf*
PRO 28: 6th International RILEM Symposium on Performance Testing and Evaluation of Bituminous Materials—PTEBM'03 (ISBN: 2-912143-35-7; e-ISBN: 978-2-912143-77-8); *Ed. M. N. Partl*
PRO 29: 2nd International RILEM Workshop on Life Prediction and Ageing Management of Concrete Structures (ISBN: 2-912143-36-5); *Ed. D. J. Naus*
PRO 30: 4th International RILEM Workshop on High Performance Fiber Reinforced Cement Composites—HPFRCC 4 (ISBN: 2-912143-37-3); *Eds. A. E. Naaman and H. W. Reinhardt*
PRO 31: International RILEM Workshop on Test and Design Methods for Steel Fibre Reinforced Concrete: Background and Experiences (ISBN: 2-912143-38-1); *Eds. B. Schnütgen and L. Vandewalle*
PRO 32: International Conference on Advances in Concrete and Structures 2 vol. (ISBN (set): 2-912143-41-1); *Eds. Ying-shu Yuan, Surendra P. Shah and Heng-lin Lü*

PRO 33: 3rd International Symposium on Self-Compacting Concrete (ISBN: 2-912143-42-X); *Eds. Ó. Wallevik and I. Níelsson*

PRO 34: International RILEM Conference on Microbial Impact on Building Materials (ISBN: 2-912143-43-8); *Ed. M. Ribas Silva*

PRO 35: International RILEM TC 186-ISA on Internal Sulfate Attack and Delayed Ettringite Formation (ISBN: 2-912143-44-6); *Eds. K. Scrivener and J. Skalny*

PRO 36: International RILEM Symposium on Concrete Science and Engineering – A Tribute to Arnon Bentur (ISBN: 2-912143-46-2); *Eds. K. Kovler, J. Marchand, S. Mindess and J. Weiss*

PRO 37: 5th International RILEM Conference on Cracking in Pavements – Mitigation, Risk Assessment and Prevention (ISBN: 2-912143-47-0); *Eds. C. Petit, I. Al-Qadi and A. Millien*

PRO 38: 3rd International RILEM Workshop on Testing and Modelling the Chloride Ingress into Concrete (ISBN: 2-912143-48-9); *Eds. C. Andrade and J. Kropp*

PRO 39: 6th International RILEM Symposium on Fibre-Reinforced Concretes—BEFIB 2004 (ISBN: 2-912143-51-9); *Eds. M. Di Prisco, R. Felicetti and G. A. Plizzari*

PRO 40: International RILEM Conference on the Use of Recycled Materials in Buildings and Structures (ISBN: 2-912143-52-7); *Eds. E. Vázquez, Ch. F. Hendriks and G. M. T. Janssen*

PRO 41: RILEM International Symposium on Environment-Conscious Materials and Systems for Sustainable Development (ISBN: 2-912143-55-1); *Eds. N. Kashino and Y. Ohama*

PRO 42: SCC'2005—China: 1st International Symposium on Design, Performance and Use of Self-Consolidating Concrete (ISBN: 2-912143-61-6); *Eds. Zhiwu Yu, Caijun Shi, Kamal Henri Khayat and Youjun Xie*

PRO 43: International RILEM Workshop on Bonded Concrete Overlays (e-ISBN: 2-912143-83-7); *Eds. J. L. Granju and J. Silfwerbrand*

PRO 44: 2nd International RILEM Workshop on Microbial Impacts on Building Materials (CD11) (e-ISBN: 2-912143-84-5); *Ed. M. Ribas Silva*

PRO 45: 2nd International Symposium on Nanotechnology in Construction, Bilbao (ISBN: 2-912143-87-X); *Eds. Peter J. M. Bartos, Yolanda de Miguel and Antonio Porro*

PRO 46: ConcreteLife'06—International RILEM-JCI Seminar on Concrete Durability and Service Life Planning: Curing, Crack Control, Performance in Harsh Environments (ISBN: 2-912143-89-6); *Ed. K. Kovler*

PRO 47: International RILEM Workshop on Performance Based Evaluation and Indicators for Concrete Durability (ISBN: 978-2-912143-95-2); *Eds. V. Baroghel-Bouny, C. Andrade, R. Torrent and K. Scrivener*

PRO 48: 1st International RILEM Symposium on Advances in Concrete through Science and Engineering (e-ISBN: 2-912143-92-6); *Eds. J. Weiss, K. Kovler, J. Marchand, and S. Mindess*

PRO 49: International RILEM Workshop on High Performance Fiber Reinforced Cementitious Composites in Structural Applications (ISBN: 2-912143-93-4); *Eds. G. Fischer and V.C. Li*

PRO 50: 1st International RILEM Symposium on Textile Reinforced Concrete (ISBN: 2-912143-97-7); *Eds. Josef Hegger, Wolfgang Brameshuber and Norbert Will*

PRO 51: 2nd International Symposium on Advances in Concrete through Science and Engineering (ISBN: 2-35158-003-6; e-ISBN: 2-35158-002-8); *Eds. J. Marchand, B. Bissonnette, R. Gagné, M. Jolin and F. Paradis*

PRO 52: Volume Changes of Hardening Concrete: Testing and Mitigation (ISBN: 2-35158-004-4; e-ISBN: 2-35158-005-2); *Eds. O. M. Jensen, P. Lura and K. Kovler*

PRO 53: High Performance Fiber Reinforced Cement Composites—HPFRCC5 (ISBN: 978-2-35158-046-2); *Eds. H. W. Reinhardt and A. E. Naaman*

PRO 54: 5th International RILEM Symposium on Self-Compacting Concrete (ISBN: 978-2-35158-047-9); *Eds. G. De Schutter and V. Boel*

PRO 55: International RILEM Symposium Photocatalysis, Environment and Construction Materials (ISBN: 978-2-35158-056-1); *Eds. P. Baglioni and L. Cassar*

PRO 56: International RILEM Workshop on Integral Service Life Modelling of Concrete Structures (ISBN 978-2-35158-058-5); *Eds. R. M. Ferreira, J. Gulikers and C. Andrade*

PRO 57: RILEM Workshop on Performance of cement-based materials in aggressive aqueous environments (e-ISBN: 978-2-35158-059-2); *Ed. N. De Belie*

PRO 58: International RILEM Symposium on Concrete Modelling—CONMOD'08 (ISBN: 978-2-35158-060-8); *Eds. E. Schlangen and G. De Schutter*

PRO 59: International RILEM Conference on On Site Assessment of Concrete, Masonry and Timber Structures—SACoMaTiS 2008 (ISBN set: 978-2-35158-061-5); *Eds. L. Binda, M. di Prisco and R. Felicetti*

PRO 60: Seventh RILEM International Symposium on Fibre Reinforced Concrete: Design and Applications—BEFIB 2008 (ISBN: 978-2-35158-064-6); *Ed. R. Gettu*

PRO 61: 1st International Conference on Microstructure Related Durability of Cementitious Composites 2 vol., (ISBN: 978-2-35158-065-3); *Eds. W. Sun, K. van Breugel, C. Miao, G. Ye and H. Chen*

PRO 62: NSF/ RILEM Workshop: In-situ Evaluation of Historic Wood and Masonry Structures (e-ISBN: 978-2-35158-068-4); *Eds. B. Kasal, R. Anthony and M. Drdácký*

PRO 63: Concrete in Aggressive Aqueous Environments: Performance, Testing and Modelling, 2 vol., (ISBN: 978-2-35158-071-4); *Eds. M. G. Alexander and A. Bertron*

PRO 64: Long Term Performance of Cementitious Barriers and Reinforced Concrete in Nuclear Power Plants and Waste Management—NUCPERF 2009 (ISBN: 978-2-35158-072-1); *Eds. V. L'Hostis, R. Gens, C. Gallé*

PRO 65: Design Performance and Use of Self-consolidating Concrete—SCC'2009 (ISBN: 978-2-35158-073-8); *Eds. C. Shi, Z. Yu, K. H. Khayat and P. Yan*

PRO 66: 2nd International RILEM Workshop on Concrete Durability and Service Life Planning—ConcreteLife'09 (ISBN: 978-2-35158-074-5); *Ed. K. Kovler*
PRO 67: Repairs Mortars for Historic Masonry (e-ISBN: 978-2-35158-083-7); *Ed. C. Groot*
PRO 68: Proceedings of the 3rd International RILEM Symposium on 'Rheology of Cement Suspensions such as Fresh Concrete (ISBN 978-2-35158-091-2); *Eds. O. H. Wallevik, S. Kubens and S. Oesterheld*
PRO 69: 3rd International PhD Student Workshop on 'Modelling the Durability of Reinforced Concrete (ISBN: 978-2-35158-095-0); *Eds. R. M. Ferreira, J. Gulikers and C. Andrade*
PRO 70: 2nd International Conference on 'Service Life Design for Infrastructure' (ISBN set: 978-2-35158-096-7, e-ISBN: 978-2-35158-097-4); *Ed. K. van Breugel, G. Ye and Y. Yuan*
PRO 71: Advances in Civil Engineering Materials—The 50-year Teaching Anniversary of Prof. Sun Wei' (ISBN: 978-2-35158-098-1; e-ISBN: 978-2-35158-099-8); *Eds. C. Miao, G. Ye, and H. Chen*
PRO 72: First International Conference on 'Advances in Chemically-Activated Materials—CAM'2010' (2010), 264 pp., ISBN: 978-2-35158-101-8; e-ISBN: 978-2-35158-115-5, *Eds. Caijun Shi and Xiaodong Shen*
PRO 73: 2nd International Conference on 'Waste Engineering and Management—ICWEM 2010' (2010), 894 pp., ISBN: 978-2-35158-102-5; e-ISBN: 978-2-35158-103-2, *Eds. J. Zh. Xiao, Y. Zhang, M. S. Cheung and R. Chu*
PRO 74: International RILEM Conference on 'Use of Superabsorsorbent Polymers and Other New Addditives in Concrete' (2010) 374 pp., ISBN: 978-2-35158-104-9; e-ISBN: 978-2-35158-105-6; *Eds. O.M. Jensen, M.T. Hasholt, and S. Laustsen*
PRO 75: International Conference on 'Material Science—2nd ICTRC—Textile Reinforced Concrete—Theme 1' (2010) 436 pp., ISBN: 978-2-35158-106-3; e-ISBN: 978-2-35158-107-0; *Ed. W. Brameshuber*
PRO 76: International Conference on 'Material Science—HetMat—Modelling of Heterogeneous Materials—Theme 2' (2010) 255 pp., ISBN: 978-2-35158-108-7; e-ISBN: 978-2-35158-109-4; *Ed. W. Brameshuber*
PRO 77: International Conference on 'Material Science—AdIPoC—Additions Improving Properties of Concrete—Theme 3' (2010) 459 pp., ISBN: 978-2-35158-110-0; e-ISBN: 978-2-35158-111-7; *Ed. W. Brameshuber*
PRO 78: 2nd Historic Mortars Conference and RILEM TC 203-RHM Final Workshop—HMC2010 (2010) 1416 pp., e-ISBN: 978-2-35158-112-4; *Eds J. Válek, C. Groot, and J. J. Hughes*
PRO 79: International RILEM Conference on Advances in Construction Materials Through Science and Engineering (2011) 213 pp., e-ISBN: 978-2-35158-117-9; *Eds Christopher Leung and K.T. Wan*
PRO 80: 2nd International RILEM Conference on Concrete Spalling due to Fire Exposure (2011) 453 pp., ISBN: 978-2-35158-118-6, e-ISBN: 978-2-35158-119-3; *Eds E.A.B. Koenders and F. Dehn*

PRO 81: 2nd International RILEM Conference on Strain Hardening Cementitious Composites (SHCC2-Rio) (2011) 451 pp., ISBN: 978-2-35158-120-9, e-ISBN: 978-2-35158-121-6; *Eds. R.D. Toledo Filho, F.A. Silva, E.A.B. Koenders and E.M. R. Fairbairn*

PRO 82: 2nd International RILEM Conference on Progress of Recycling in the Built Environment (2011) 507 pp., e-ISBN: 978-2-35158-122-3; *Eds. V.M. John, E. Vazquez, S.C. Angulo and C. Ulsen*

PRO 83: 2nd International Conference on Microstructural-related Durability of Cementitious Composites (2012) 250 pp., ISBN: 978-2-35158-129-2; e-ISBN: 978-2-35158-123-0; *Eds. G. Ye, K. van Breugel, W. Sun and C. Miao*

PRO 85: RILEM-JCI International Workshop on Crack Control of Mass Concrete and Related issues concerning Early-Age of Concrete Structures—ConCrack 3—Control of Cracking in Concrete Structures 3 (2012) 237 pp., ISBN: 978-2-35158-125-4; e-ISBN: 978-2-35158-126-1; *Eds F. Toutlemonde and J.-M. Torrenti*

PRO 86: International Symposium on Life Cycle Assessment and Construction (2012) 414 pp., ISBN: 978-2-35158-127-8, e-ISBN: 978-2-35158-128-5; *Eds A. Ventura and C. de la Roche*

PRO 87: UHPFRC 2013—RILEM-fib-AFGC International Symposium on Ultra-High Performance Fibre-Reinforced Concrete (2013), ISBN: 978-2-35158-130-8, e-ISBN: 978-2-35158-131-5; *Eds F. Toutlemonde*

PRO 88: 8th RILEM International Symposium on Fibre Reinforced Concrete (2012) 344 pp., ISBN: 978-2-35158-132-2, e-ISBN: 978-2-35158-133-9; *Eds Joaquim A.O. Barros*

PRO 89: RILEM International workshop on performance-based specification and control of concrete durability (2014) 678 pp, ISBN: 978-2-35158-135-3, e-ISBN: 978-2-35158-136-0; *Eds. D. Bjegović, H. Beushausen and M. Serdar*

PRO 90: 7th RILEM International Conference on Self-Compacting Concrete and of the 1st RILEM International Conference on Rheology and Processing of Construction Materials (2013) 396 pp, ISBN: 978-2-35158-137-7, e-ISBN: 978-2-35158-138-4, *Eds. Nicolas Roussel and Hela Bessaies-Bey*

PRO 91 draft: CONMOD 2014—RILEM International Symposium on Concrete Modelling (2014), ISBN: 978-2-35158-139-1; e-ISBN: 978-2-35158-140-7

PRO 92: CAM 2014—2nd International Conference on advances in chemically-activated materials (2014) 392 pp., ISBN: 978-2-35158-141-4; e-ISBN: 978-2-35158-142-1, *Eds. Caijun Shi and Xiadong Shen*

PRO 93: SCC 2014—3rd International Symposium on Design, Performance and Use of Self-Consolidating Concrete (2014) 438 pp., ISBN: 978-2-35158-143-8; e-ISBN: 978-2-35158-144-5, *Eds. Caijun Shi, Zhihua Ou, Kamal H. Khayat*

PRO 94 (online version): HPFRCC-7—7th RILEM conference on High performance fiber reinforced cement composites, e-ISBN: 978-2-35158-146-9, *Eds. H.W. Reinhardt, G.J. Parra-Montesinos, H. Garrecht*

PRO 95: International RILEM Conference on Application of superabsorbent polymers and other new admixtures in concrete construction, ISBN: 978-2-35158-147-6; e-ISBN: 978-2-35158-148-3, *Eds. Viktor Mechtcherine, Christof Schroefl*

PRO 96 (online version): XIII DBMC: XIII International Conference on Durability of Building Materials and Components, e-ISBN: 978-2-35158-149-0, *Eds. M. Quattrone, V.M. John*

PRO 97: SHCC3—3rd International RILEM Conference on Strain Hardening Cementitious Composites, ISBN: 978-2-35158-150-6; e-ISBN: 978-2-35158-151-3, *Eds. E. Schlangen, M.G. Sierra Beltran, M. Lukovic, G. Ye*

PRO 98: FERRO-11—11th International Symposium on Ferrocement and 3rd ICTRC—International Conference on Textile Reinforced Concrete, ISBN: 978-2-35158-152-0; e-ISBN: 978-2-35158-153-7, *Ed. W. Brameshuber*

PRO 99 (online version): ICBBM 2015—1st International Conference on Bio-Based Building Materials, e-ISBN: 978-2-35158-154-4, *Eds. S. Amziane, M. Sonebi*

PRO 100: SCC16—RILEM Self-Consolidating Concrete Conference, ISBN: 978-2-35158-156-8; e-ISBN: 978-2-35158-157-5

PRO 101 (online version): III Progress of Recycling in the Built Environment, e-ISBN: 978-2-35158-158-2, *Eds. M. Quattrone, V.M. John*

RILEM Reports (REP)

Report 19: Considerations for Use in Managing the Aging of Nuclear Power Plant Concrete Structures (ISBN: 2-912143-07-1); *Ed. D. J. Naus*

Report 20: Engineering and Transport Properties of the Interfacial Transition Zone in Cementitious Composites (ISBN: 2-912143-08-X); *Eds. M. G. Alexander, G. Arliguie, G. Ballivy, A. Bentur and J. Marchand*

Report 21: Durability of Building Sealants (ISBN: 2-912143-12-8); *Ed. A. T. Wolf*

Report 22: Sustainable Raw Materials—Construction and Demolition Waste (ISBN: 2-912143-17-9); *Eds. C. F. Hendriks and H. S. Pietersen*

Report 23: Self-Compacting Concrete state-of-the-art report (ISBN: 2-912143-23-3); *Eds. Å. Skarendahl and Ö. Petersson*

Report 24: Workability and Rheology of Fresh Concrete: Compendium of Tests (ISBN: 2-912143-32-2); *Eds. P. J. M. Bartos, M. Sonebi and A. K. Tamimi*

Report 25: Early Age Cracking in Cementitious Systems (ISBN: 2-912143-33-0); *Ed. A. Bentur*

Report 26: Towards Sustainable Roofing (Joint Committee CIB/RILEM) (CD 07) (e-ISBN 978-2-912143-65-5); *Eds. Thomas W. Hutchinson and Keith Roberts*

Report 27: Condition Assessment of Roofs (Joint Committee CIB/RILEM) (CD 08) (e-ISBN 978-2-912143-66-2); *Ed. CIB W 83/RILEM TC166-RMS*

Report 28: Final report of RILEM TC 167-COM 'Characterisation of Old Mortars with Respect to Their Repair (ISBN: 978-2-912143-56-3); *Eds. C. Groot, G. Ashall and J. Hughes*
Report 29: Pavement Performance Prediction and Evaluation (PPPE): Interlaboratory Tests (e-ISBN: 2-912143-68-3); *Eds. M. Partl and H. Piber*
Report 30: Final Report of RILEM TC 198-URM 'Use of Recycled Materials' (ISBN: 2-912143-82-9; e-ISBN: 2-912143-69-1); *Eds. Ch. F. Hendriks, G. M. T. Janssen and E. Vázquez*
Report 31: Final Report of RILEM TC 185-ATC 'Advanced testing of cement-based materials during setting and hardening' (ISBN: 2-912143-81-0; e-ISBN: 2-912143-70-5); *Eds. H. W. Reinhardt and C. U. Grosse*
Report 32: Probabilistic Assessment of Existing Structures. A JCSS publication (ISBN 2-912143-24-1); *Ed. D. Diamantidis*
Report 33: State-of-the-Art Report of RILEM Technical Committee TC 184-IFE 'Industrial Floors' (ISBN 2-35158-006-0); *Ed. P. Seidler*
Report 34: Report of RILEM Technical Committee TC 147-FMB 'Fracture mechanics applications to anchorage and bond' Tension of Reinforced Concrete Prisms – Round Robin Analysis and Tests on Bond (e-ISBN 2-912143-91-8); *Eds. L. Elfgren and K. Noghabai*
Report 35: Final Report of RILEM Technical Committee TC 188-CSC 'Casting of Self Compacting Concrete' (ISBN 2-35158-001-X; e-ISBN: 2-912143-98-5); *Eds. Å. Skarendahl and P. Billberg*
Report 36: State-of-the-Art Report of RILEM Technical Committee TC 201-TRC 'Textile Reinforced Concrete' (ISBN 2-912143-99-3); *Ed. W. Brameshuber*
Report 37: State-of-the-Art Report of RILEM Technical Committee TC 192-ECM 'Environment-conscious construction materials and systems' (ISBN: 978-2-35158-053-0); *Eds. N. Kashino, D. Van Gemert and K. Imamoto*
Report 38: State-of-the-Art Report of RILEM Technical Committee TC 205-DSC 'Durability of Self-Compacting Concrete' (ISBN: 978-2-35158-048-6); *Eds. G. De Schutter and K. Audenaert*
Report 39: Final Report of RILEM Technical Committee TC 187-SOC 'Experimental determination of the stress-crack opening curve for concrete in tension' (ISBN 978-2-35158-049-3); *Ed. J. Planas*
Report 40: State-of-the-Art Report of RILEM Technical Committee TC 189-NEC 'Non-Destructive Evaluation of the Penetrability and Thickness of the Concrete Cover' (ISBN 978-2-35158-054-7); *Eds. R. Torrent and L. Fernández Luco*
Report 41: State-of-the-Art Report of RILEM Technical Committee TC 196-ICC 'Internal Curing of Concrete' (ISBN 978-2-35158-009-7); *Eds. K. Kovler and O. M. Jensen*
Report 42: 'Acoustic Emission and Related Non-destructive Evaluation Techniques for Crack Detection and Damage Evaluation in Concrete'—Final Report of RILEM Technical Committee 212-ACD (e-ISBN: 978-2-35158-100-1); *Ed. M. Ohtsu*

RILEM State-of-the-Art Reports (STARs)

Volume 1: State-of-the-Art Report of the RILEM Technical Committee 207-INR 'Non-Destructive Assessment of Concrete Structures: Reliability and Limits of Single and Combined Techniques' (ISBN: 978-94-007-2735-9; e-ISBN: 978-94-007-2736-6); *Ed. D. Breysse* (2012)
Volume 2: State-of-the-Art Report Prepared by Technical Committee 225-SAP 'Application of Super Absorbent Polymers (SAP) in Concrete Construction' (ISBN: 978-94-007-2732-8; e-ISBN: 978-94-007-2733-5); *Eds. V. Mechtcherin and H.-W. Reinhardt* (2012)
Volume 3: State-of-the-Art Report of the RILEM Technical Committee 193-RLS 'Bonded Cement-Based Material Overlays for the Repair, the Lining or the Strengthening of Slabs or Pavements' (ISBN: 978-94-007-1238-6; e-ISBN: 978-94-007-1239-3); *Eds. B. Bissonnette, L. Courard, D. W. Fowler and J.-L. Granju* (2011)
Volume 4: 'Durability of Strain-Hardening Fibre-Reinforced Cement-Based Composites (SHCC)' (ISBN: 978-94-007-0337-7; e-ISBN: 978-94-007-0338-4); *Eds. F. Wittmann and G. Van Zijl* (2011)
Volume 5: State-of-the-Art Report of the RILEM Technical Committee 194-TDP 'Applications of Titanium Dioxide Photocatalysis to Construction Materials' (ISBN: 978-94-007-1296-6; e-ISBN: 978-94-007-1297-3); *Eds. Y. Ohama and D. Van Gemert* (2011)
Volume 6: State-of-the-Art Report of the RILEM Technical Committee 208-HFC, SC3 'Strain Hardening Cement Composites: Structural Design and Performance' (ISBN: 978-94-007-4835-4; e-ISBN: 978-94-007-4836-1); *Eds. K. Rokugo and T. Kanda* (2013)
Volume 7: State of the Art Report of the RILEM Technical Committee 215-AST 'In Situ Assessment of Structural Timber' (ISBN: 978-94-007-0559-3; e-ISBN: 978-94-007-0560-9); *Eds. B. Kasal and T. Tannert* (2011)
Volume 8: Final report of the RILEM Technical Committee 217-PRE 'Progress of Recycling in the Built Environment' (ISBN: 978-94-007-4907-8; e-ISBN: 978-94-007-4908-5); *Ed. E. Vázquez* (2013)
Volume 9: State-of-the-Art Report of the RILEM Technical Committee 206-ATB 'Advances in Interlaboratory Testing and Evaluation of Bituminous Materials' (ISBN: 978-94-007-5103-3; e-ISBN: 978-94-007-5104-0); *Eds. M. N. Partl, H. U. Bahia, F. Canestrari, C. de la Roche, H. Di Benedetto, H. Piber and D. Sybilski* (2013)
Volume 10: State-of-the-Art Report, RILEM TC 211 - PAE 'Performance of Cement-Based Materials in Aggressive Aqueous Environments' (ISBN: 978-94-007-5412-6; e-ISBN: 978-94-007-5413-3); *Eds. M. Alexander, A. Bertron and N. De Belie* (2013)
Volume 11: State-of-the-Art Report of RILEM Technical Committee 221-SHC 'Self-Healing Phenomena in Cement-Based Materials' (ISBN: 978-94-007-6623-5;

e-ISBN: 978-94-007-6624-2); *Eds. M. de Rooij, K. Van Tittelboo, N. De Belie and E. Schlangen* (2013)

Volume 12: Part 1 Diagnosis (AAR 6.1) 'Guide to Diagnosis and Appraisal of AAR Damage to Concrete in Structures' (ISBN: 978-94-007-6566-5; e-ISBN: 978-94-007-6567-2); *Eds. B. Godart, M. de Rooij and J. G.M. Wood* (2013)

Volume 13: State-of-the-Art Report, RILEM TC 224-AAM 'Alkali Activated Materials' (ISBN: 978-94-007-7671-5; e-ISBN: 978-94-007-7672-2); *Eds. J. L. Provis and J. S. J. van Deventer* (2014)

Volume 14: State-of-the-Art Report of the RILEM Technical Committee 228-MPS 'Mechanical Properties of Self-Compacting Concrete' (ISBN: 978-3-319-03244-3; e-ISBN: 978-3-319-03245-0); *Eds. K. H. Khayat and G. De Schutter* (2014)

Volume 15: State-of-the Art Report of the RILEM Technical Committee 222-SCF 'Simulation of Fresh Concrete Flow' (ISBN: 978-94-017-8883-0; e-ISBN: 978-94-017-8884-7); *Eds. N. Roussel and A. Gram* (2014)

Volume 16: 'RILEM Technical Committee 195-DTD Recommendation for Test Methods for AD and TD of Early Age Concrete'. Round Robin Documentation Report: Program, Test Results and Statistical Evaluation (ISBN: 978-94-017-9265-3; e-ISBN: 978-94-017-9266-0); *Eds. Ø. Bjøntegaard, T. A. Martius-Hammer, M. Krauss and H. Budelmann* (2015)

Volume 17: State-of-the-Art Report of the RILEM Technical Committee 219-ACS 'RILEM Recommendations for the Prevention of Damage by Alkali-Aggregate Reactions in New Concrete Structures' (ISBN: 978-94-017-7251-8; e-ISBN: 978-94-017-7252-5); *Eds. Philip J. Nixon and I. Sims* (2016)

Volume 18: State-of-the-Art Report RILEM TC 230-PSC 'Performance-Based Specifications and Control of Concrete Durability' (ISBN: 978-94-017-7308-9; e-ISBN: 978-94-017-7309-6); *Eds. H. Beushausen and L. Fernandez Luco* (2016)

Volume 19: State-of-the-Art Report of the RILEM Technical Committee 234-DUC 'Design Procedures for the Use of Composites in Strengthening of Reinforced Concrete Structures' (ISBN: 978-94-017-7335-5; e-ISBN: 978-94-017-7336-2); *Eds. C. Pellegrino and J. Sena-Cruz* (2015)

Contents

Introduction

This book contains the full suite of RILEM Recommendations that have been produced to enable engineers, specifiers and testing houses to design and produce concrete that will not suffer damage arising from alkali-aggregate reactions in the concrete. There are five Recommended methods of test for aggregates (designated AAR-1 to AAR-5) and an overall Recommendation which describes how these should be used to enable a comprehensive aggregate assessment (AAR-0). Additionally, there are two Recommended International Specifications for concrete (AAR-7.1 and 7.2) and a Preliminary International Specification for dams and other hydro structures (AAR-7.3), which describe how the aggregate assessment can be combined with other measures in the design of the concrete to produce a concrete with a minimised risk of developing damage from alkali-aggregate reactions. In detail, this suite of Recommendations consists of the following:

AAR-0: Outline Guide to the Use of RILEM Methods in Assessments of Alkali-Reactivity Potential of Aggregates
AAR-1.1: Detection of Potential Alkali-Reactivity—RILEM Petrographic Examination Method
AAR-2: Detection of Potential Alkali-Reactivity—Accelerated Mortar-bar Test Method for Aggregates
AAR-3: Detection of Potential Alkali-Reactivity—38 °C Test Method for Aggregate Combinations using Concrete Prisms
AAR-4.1: Detection of Potential Alkali-Reactivity—60 °C Test Method for Aggregate Combinations using Concrete Prisms
AAR-5: Rapid Preliminary Screening Test for Carbonate Aggregates
AAR-7.1: International Specification to Minimise Damage from Alkali Reactions in Concrete; Part 1—Alkali-Silica Reaction
AAR-7.2: International Specification to Minimise Damage from Alkali Reactions in Concrete; Part 2—Alkali-Carbonate Reaction
AAR-7.3: Preliminary International Specification to Minimise Damage from Alkali Reactions in Concrete; Part 3—Concrete Dams and Other Hydro Structures

P.J. Nixon and I. Sims (eds.), *RILEM Recommendations for the Prevention of Damage by Alkali-Aggregate Reactions in New Concrete Structures*, RILEM State-of-the-Art Reports 17, DOI 10.1007/978-94-017-7252-5_1

Note: *AAR-6.1 has already been published separately, whilst AAR-1.2 is due to be published shortly and, AAR-6.2 (incorporating AAR-9) and AAR-8 are still under development and will be published separately.*

These Recommendations have been developed by three successive RILEM Technical Committees:

RILEM TC 106, Alkali Aggregate Reactions- Accelerated Tests, was set up in 1988 specifically to address the problems arising from the proliferation of test methods for aggregate reactivity. As knowledge of the mechanism of the reaction grew it became clear that AAR is a family of related reactions affecting different rock types in different ways. Accordingly, each region developed tests appropriate to its own geology and experience with the reaction. A second reason for this proliferation was the need to accelerate the reaction in any test method. Alkali reactions take many years to exhibit damaging expansion. So for a test to be useful for assessment and contractual purposes fairly drastic acceleration is necessary. This, however, almost certainly distorts the test results from the reality of field concrete and the search for a solution that satisfies both the desire for a speedy result and the requirement that laboratory findings should reliably reflect field experience has also generated a multitude of methods. It was against this background that RILEM TC-106 was formed with a primary objective to recommend a suite of test methods that would be:

- reliable;
- closely related to field experience;
- sufficiently accelerated to be practically useful;
- internationally accepted.

After considering a wide range of methods for the assessment of aggregates for alkali-silica reactivity, TC 106 initially concentrated upon three procedures: petrographic examination (AAR-1), an accelerated mortar-bar expansion test (AAR-2) and a concrete prism expansion test (AAR-3). AAR-2 and AAR-3 were published in 2000 as draft RILEM Recommendations and considerable progress was made in developing AAR-1. The results of AAR-2 and AAR-3 were assessed against the known field performance of a wide range of aggregates, of different geological types and in different countries, in an international inter-laboratory trial, and to improve the reliability of the methods, sources of reference high-alkali cements and non-reactive aggregates were established. Work was also started by TC-106 on an accelerated concrete prism test (AAR-4) and on specialised procedures for carbonate aggregates.

The succeeding TC, RILEM TC 191-ARP, Alkali Reactivity; Prevention, Assessment, Specification and Diagnosis, began work in 2000 with terms of reference to develop integrated assessment, specification and diagnosis procedures for alkali-reactions in concrete, building on the work of TC-106. It set out to integrate the test methods already developed by TC-106 into a coherent assessment system. Additionally it began work on guidance on the reliable diagnosis of damage from alkali-reactions (AAR-6) and on the principles and options for the specification of concrete to avoid such damage (AAR-7). Under the aegis of TC 191-ARP, AAR-1 was published as a draft RILEM recommendation and the AAR-4 procedure was finalised and an international inter-laboratory trial carried out. Also, the preliminary screening method for carbonate aggregates (AAR-5) was developed, assessed in an inter-laboratory trial then published in Materials and Structures as a draft RILEM Recommendation and the two International Specifications, AAR 7.1 and 7.2, were developed and presented for discussion at International Conferences.

During this period an EU funded project, PARTNER, was also undertaken to assess the RILEM methods as the basis for European (CEN) Standard methods. In this project 24 partners from 14 European countries evaluated the RILEM methods using 22 different aggregates. Some regional methods were also evaluated and field trials with large concrete specimens established for long term tests. Broadly, this project showed that the RILEM methods could successfully identify the reactivity of the aggregates tested. The work also included an inter-laboratory precision test and recommendations for improvements in the procedures were made.

Under the third TC, RILEM TC 219-ACS, Alkali-Aggregate Reactions in Concrete and Structures, the lessons of the PARTNER programme were incorporated into the RILEM methods, the Specifications were finalised and the entire suite of methods and specifications integrated, reassessed, updated and prepared for this publication. Additionally, a worldwide petrographic atlas of reactive aggregates has been produced to assist in the examination of aggregates using AAR-1.1. This will be published separately as RILEM AAR-1.2, and hopefully also, in due course, a guidance document giving further detailed information on the methodology of petrographic examination and supplementary methods for establishing the composition and texture of aggregates.

The work of TC 219-ACS was especially extended to include the development of a performance test for specific concrete mixes, the provision of guidance on the appraisal, management and modelling of structures affected by damaging AAR (AAR-6.2 and AAR-9) and the development of a method (AAR-8) for assessing the additional reactive alkalis that can be leached from aggregates. These items are ongoing and are not covered by this publication.

TC 219-ACS formally completed its programme in early 2014. A new committee has been formed, TC AAA, under the chairmanship of Prof. Borge Wigum, to carry on the work, with particular references to the performance test development, long-term exposure sites for assistance with validating tests, and methods for assessing the long-term availability of alkalis within concrete. TC AAA held its inaugural meeting in autumn 2014.

During the period of these various TCs, a much better international consensus has developed and some of the RILEM methods, including AAR-1, 2, 3 and 4, are now used routinely in many countries (some of which have incorporated similar methods into their national standards). Nevertheless, novel methods are still being developed and promoted and the need to achieve and maintain an international consensus has not gone away. It is believed that these RILEM Recommendations form the basis for such a successful international consensus.

Philip Nixon, Chairman, RILEM TC 219-ACS
Ian Sims, Secretary, RILEM TC 219-ACS
May 2014 (updated March 2015)

The final membership of RILEM TC 219-ACS was as follows, including some retired members from the previous TC 106-AAR and TC 191-ARP (in alphabetical order):

Andiç-Çakir, O (Turkey)	Lindgård, J (Norway)[c]
Appelquist, K (Sweden)	López Buendia, A (Spain)
Berra, M (Italy)	Lu, D-Y (China)
Bérubé, M-A (Canada)	Marfil, S (Argentina)
Borchers, I (Germany)	Marion, A-M (Belgium)
Bremseth, SK (Norway)	Martin, R-P (France)
Broekmans, MATM (Netherlands and Norway)[a]	Menendez Mendez, E (Spain)[a]
Brouard, E (France)[a]	Merz, C (Switzerland)
Capra, B (France)	Mueller, U (Sweden)
Döse, M (Sweden)	Müller, C (Germany)
Dunant, C (Switzerland)	Mullick, AK (India)
Eriksen, K (Denmark)	Nixon, PJ (UK): Chairman
Fairbairn, E (Brazil)	Pederson, B (Norway)
Fernandes, I (Portugal)[a]	Pierre, C (Belgium)
Folliard, KJ (USA)	Ramezanianpour, AA (Iran)
Fournier, B (Canada)	Rogers, CA (Canada)
Freitag, S (New Zealand)	Rønning, TF (Norway)[a]
Frigerio, A (Italy)	de Rooij, M (Netherlands)
Giebson, C (Germany)	Santos Silva, A (Portugal)
Godart, B (France)[a]	Sauvaget, C (France)
Gomes, JP (Portugal)	Schouenborg, B (Sweden)
Góralczyk, S (Poland)	Scrivener, K (UK and Switzerland)
Grattan-Bellew, PE (Canada)	Seignol, J-F (France)[a]
de Grosbois, M (Canada)	Seyfarth, K (Germany)
Haugen, M (Norway)	Shayan, A (Australia)
Holt, E (USA and Finland)	Sims, I (UK): Secretary
Hooton, RD (Canada)	Sommer, H (Austria)[a]
Jensen, V (Denmark and Norway)[a]	Steigenberger, J (Austria)
Justnes, H (Norway)[a]	Stemland, H (Norway)
Kara, P (Latvia)	Sujjavnich, S (Thailand)
Katayama, T (Japan)	Tang, M-S (China)
Krispel, S (Austria)	Thomas, MDA (UK and Canada)
Kuperman, SC (Brazil)	Vaičienė, M (Lithuania)
Lagerblad, B (Sweden)	Weise, F (Germany)
Larive, C (France)	Wigum, JF (Norway and Iceland)[b]
Leemann, A (Switzerland)	Winnicki, A (Poland)
Leung, P (Hong Kong, China)	Wood, JGM (UK)

[a]Team leaders
[b]Chairman of follow-on TC AAA
[c]Secretary of follow-on TC AAA

RILEM Recommended Test Method: AAR-0—Outline Guide to the Use of RILEM Methods in the Assessment of the Alkali-Reactivity Potential of Aggregates

1 Introduction

AAR-0 provides guidance on the integrated use of the assessment procedures described in AAR-1.1 & 1.2, AAR-2, AAR-3, AAR-4.1 and AAR-5 including preliminary advice on the interpretation of their findings. The principles are illustrated by the flow chart given in Fig. 1. Guidance on the specialised assessment of carbonate rock aggregates for alkali-reactivity potential is given in Annex A. Information on reference materials and testing accessories is given in Annex B.

2 Aggregate Assessment

Aggregates from both new and existing sources frequently require to be assessed for their suitability for use in concrete. The investigation of AAR potential is one essential part of the assessment, but it should be recognised that, in many or most cases, other properties will have a more important potential influence on the performance and durability of aggregates. Therefore, the evaluation of AAR potential should not be carried out in isolation, but rather as a specialised extension to the routine assessment of the suitability of an aggregate.

Consideration of AAR potential is complicated by the so-called 'pessimum' behaviour of some aggregates, whereby expansion of concrete is maximised at a certain level of reactive constituent in the aggregate and progressively reduced for both greater and lesser levels. It is consequently important for AAR assessment to consider the total combination of coarse and fine aggregates, rather than only the individual materials.

P.J. Nixon and I. Sims (eds.), *RILEM Recommendations for the Prevention of Damage by Alkali-Aggregate Reactions in New Concrete Structures*, RILEM State-of-the-Art Reports 17, DOI 10.1007/978-94-017-7252-5_2

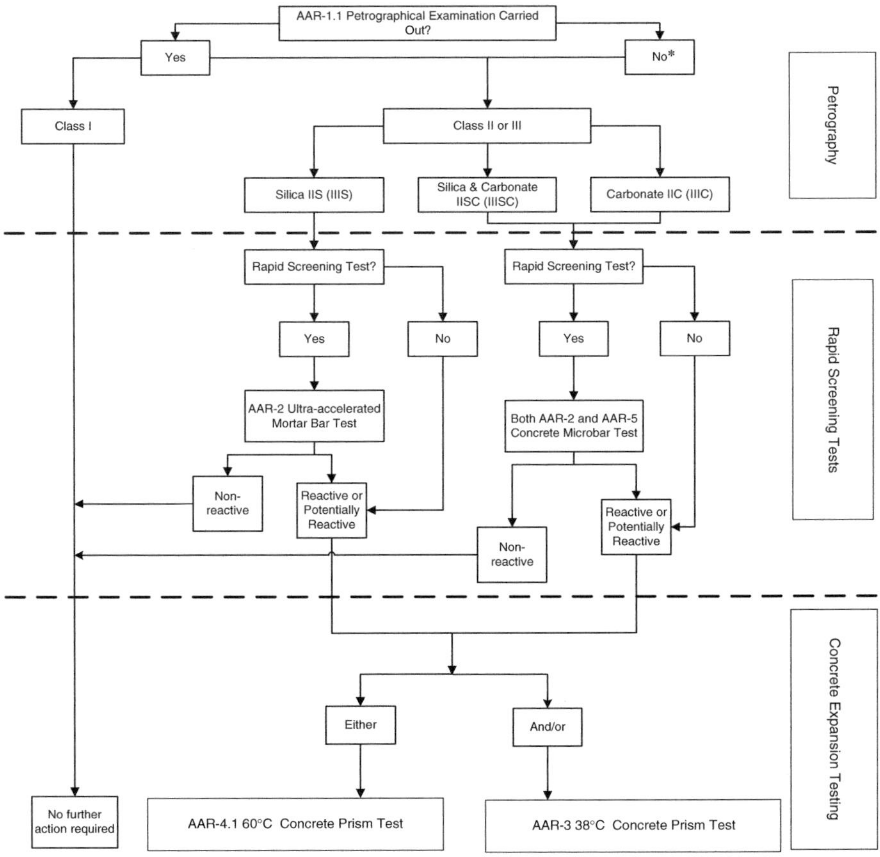

Fig. 1 Integrated assessment scheme. * If no petrographical examination has been carried out, assume Class II (or III)

Any expansion exceeding the criteria given for the tests used in the assessment of AAR potential are taken possibly to be caused by ASR and/or by reactions involving carbonates. However, it is recommended that post-test petrographical examination of specimens should be carried out to confirm that any expansion was caused by a form of AAR.

In addition to inherent reactivity, some aggregates, irrespective of whether they are themselves reactive or non-reactive, can influence the reactivity potential of a concrete mix by releasing alkalis that are additional to those derived primarily from the cement. The AAR-8 procedure for determining any content of releasable alkalis in aggregates is still under development and so is not part of the assessment scheme, but when it is available it should be carried out as part of an estimation of reactive alkali content in a particular concrete mix.

3 Principle

Any assessment of an aggregate combination for AAR potential should ideally commence with petrographical examinations of the component aggregates, which establishes their individual and combined compositions and identifies the types and concentrations of any potentially reactive constituents. This usually allows an aggregate combination to be assigned to one of three categories, as follows:

Class I—very unlikely to be alkali-reactive
Class II—potentially alkali-reactive or alkali-reactivity uncertain
Class III—very likely to be alkali-reactive

In the case of new aggregate sources, Class II is common and further testing will be required. For existing aggregate sources, when experience of use can be taken into account for local applications, Classes I or III are more often possibilities. Class III is exceptional for new aggregates and essentially limited to those found to contain opal or opaline silica.

When petrography indicates Class II (or Class III), it becomes necessary to decide on the most appropriate further tests. Aggregates which are either mainly siliceous, or carbonates with a potentially reactive silica content, are designated Class II-S or III-S and may be subjected to the RILEM expansion tests (AAR-2, AAR-3 or AAR-4.1).

Aggregates which are either mainly carbonate, or mixtures including potentially reactive types of carbonate, are designated Class II-C or III-C and may be subjected to the specialised procedures for aggregates comprising or containing carbonate materials, especially if the carbonate includes the mineral dolomite (calcium-magnesium carbonate). Some aggregates of mixed composition might be designated Class II-SC or III-SC and should thus be subjected to the procedures described for carbonate aggregates (AAR-5, but also see the guidance provided in Annex A).

The proportion of silica that can lead to the most damaging reaction will depend on the reactivity of the silica. A small amount of highly reactive silica in the aggregate will be most damaging, whereas, if the aggregate contains a high proportion of such highly reactive silica, there may be little damage. If aggregate containing highly reactive silica is mixed with a non-reactive aggregate, the behaviour of the mix will vary from very damaging to not damaging at all, depending on the proportions of the mix. This feature is known as the 'pessimum' effect. Conversely, in aggregates containing low reactivity forms of silica or where the silica is not easily exposed to the alkaline pore solution, the worst damage may occur when the greatest amount of silica is present.

Because of this it is important that the whole aggregate combination is assessed as amounts of reactive silica that are innocuous in either the fine or coarse aggregate alone may be damaging in the combined aggregate. Conversely, apparently reactive fine or coarse aggregates may be safe when used in combination. Both AAR-3 and AAR-4.1 are suitable for assessing the combined aggregate.

In the case of the RILEM expansion tests, the AAR-3 concrete prism method has previously been regarded as the reference test, on the basis of accumulated experience of its use in various forms. However, AAR-3 requires a lengthy period, up to 12 months or more, for reliable results to be obtained and even AAR-4.1 requires up to 4 months. Consequently, the accelerated mortar-bar (AAR-2) and concrete-bar (AAR-5) tests have been developed for the optional provision of an earlier indication of the outcome.

At present, following petrographic assessment, it is considered unwise to rely solely on the results of the accelerated screening tests and the preliminary indications from those methods should always be confirmed by one of the concrete prism tests. Also, practical experience has suggested that the accelerated mortar-bar test (AAR-2) might be unreliable for Class II-S aggregates containing porous flint (a type of chert) as a potentially reactive constituent. Greater experience with the accelerated mortar-bar test may, in due course, enable this advice to be modified.

It is hoped that AAR-3, or a variant of it, might eventually also be usable for assessing the reactivity performance of particular concrete mixes and the RILEM committee is actively developing such a performance test. However, although preliminary indications are encouraging, a definite correlation between the short-term results of this test method and long-term field performance has not yet been demonstrated, so that guidance on its use in practice cannot be provided at present. It is hoped that further development and international trials might, in due course, enable the performance variant of this method to be used for acceptance testing on a project-by-project basis.

All sources of natural aggregates exhibit both systematic and random variations in composition and properties. Suitability assessments have therefore to be repeated periodically and this is particularly the case with evaluations of AAR potential.

4 Samples

Laboratory investigations are only reliable if the samples are representative. It is therefore important to ensure that the sample used for AAR assessment is properly representative of its source. In the case of an operating existing quarry, it is usually appropriate to take samples from the current stockpiles of processed aggregates, following the sampling procedures given in national and international standards for aggregate testing.

In the case of a new or prospective quarry, it might be more appropriate for an experienced geologist to take rock lump samples directly from natural outcrops and/or to drill cores from rock bodies to be extracted as quarrying for aggregates proceeds. Different rock types would be tested separately or in controlled combinations at the discretion of the field geologist: the test samples should endeavour to represent the aggregates which will be produced for actual use.

Guidance on the taking of representative samples is included in AAR-1.1 (petrographical examination).

RILEM have established some sources of suitable reference materials, including high-alkali Portland cement and both reactive and non-reactive natural aggregates (see Annex B).

5 Petrographical Examination: AAR-1.1 & 1.2

A procedure is given in AAR-1.1 & 1.2 for the petrographical examination and classification of aggregate samples for AAR potential. This procedure enables any potentially alkali-reactive constituents to be identified and, if necessary, quantified. The identification is based primarily upon basic petrological or mineralogical type (s), supported, whenever possible and appropriate, by local experience.

As explained earlier, petrographical examination will lead to one of three Classes: I, II or III. In the case of Class II (or Class III), it will also be necessary for the petrographical examination to determine whether the aggregate is wholly or partly siliceous (Class II-S or III-S), or wholly or partly carbonate (Class II-C or III-C), or possibly a combination containing significant proportions of both siliceous and carbonate materials (Class II-SC or III-SC). If petrography is not available or was inconclusive, the material being evaluated should be regarded as being Class II.

The procedure described in AAR-1.1 & 1.2 results in a petrographic analysis for the sample under investigation, whereby each particulate constituent has been petrologically (or mineralogically) identified, its relative proportion determined and its alkali-reactivity status (judged innocuous or potentially reactive) established. This information is then used to classify the aggregate sample, for the purposes of the AAR assessment, into one of the three categories I, II or III, suffixed -S, -C or -SC as appropriate.

Acceptance and experience with reactive constituents differ between countries, and thus, final assessment and classification should follow any national or regional experiences, recommendations and specifications. Therefore, it is recommended that, whenever possible, petrographers should apply local guidance and/or local experience to assist with this classification.

In the case of Class II and III aggregate samples, the material is additionally sub-classified according to the siliceous and/or carbonate nature of the potentially reactive constituents, using the following definitions:

Classes II-S & III-S aggregate samples contain particulate constituents judged to be potentially alkali-silica reactive (ASR).
Classes II-C & III-C aggregate samples contain particulate carbonate constituents judged to be potentially reactive.
Classes II-SC & III-SC aggregate samples contain both particulate constituents judged to be potentially alkali-silica reactive (ASR) and particulate carbonate constituents judged to be potentially reactive.

In the case of Class II-S or III-S materials, it is then appropriate to carry out the RILEM test methods for alkali-silica reactivity (ASR): the accelerated mortar-bar test, AAR-2, for short-term screening purposes and the 38 °C concrete prism test, AAR-3, for any long-term confirmation. The 60 °C accelerated concrete prism test, AAR-4.1, may be considered as an alternative to AAR-3.

In the case of Class II-C, II-SC, III-C or III-SC materials, it is instead appropriate to carry out the AAR-5 short-term screening test procedures for aggregates comprising or containing carbonate aggregates. Again, any long-term confirmatory testing will involve either or both of the AAR-4.1 and AAR-3 methods. Additional information on the assessment of carbonate rock aggregates is given in Annex A.

Practical experience has indicated, however, that Class II-S or III-S aggregates containing more than 2 % by mass porous flint (chert) as a potentially reactive constituent cannot be reliably assessed using the AAR-2 accelerated mortar-bar test. Such aggregates are widely encountered, for example, in several northern European countries, including Belgium, Denmark, the Netherlands and the United Kingdom. Some porous flint (chert) aggregate combinations that have been established as being expansively reactive in actual structures were not detected as being expansive in the accelerated mortar-bar test. Class II-S or III-S aggregates found by petrography to contain more than 2 % porous flint (chert), therefore, should either be assessed using the AAR-3 or AAR-4.1 concrete prism tests or accepted as being potentially alkali-reactive and precautions taken to minimise the risk of ASR damage to any concrete in which the material is used.

6 Accelerated Mortar-Bar Testing: AAR-2

An accelerated screening test for ASR, using mortar-bar specimens, is given in AAR-2. The method is unsuitable for porous flint (chert) aggregates (see above).

Experience has shown that the test procedure is able to detect pessimum behaviour and it is therefore recommended that a series of tests is carried out, in which the test aggregate is mixed with a non-reactive material in a range of proportions. Guidance on this procedure is given in the annex to AAR-2. However, it is not certain that the pessimum proportion indicated by the test corresponds with that exhibited by a comparable concrete.

Criteria for the interpretation of the results of AAR-2 have not yet been finally agreed. However, on the basis of trials carried out by RILEM on aggregate combinations of known field performance from various parts of the world, it seems that results in the test (after the standard 16-days, using 'long thin' 25 × 25 × 250–300 mm specimens) of less than 0.10 % are likely to indicate non-expansive materials, whilst results exceeding 0.20 % are likely to indicate expansive materials. It is not currently possible to provide definitive interpretative guidance for results in the intermediate range 0.10–0.20 % and, for all practical purposes in the absence of additional local experience, aggregates yielding AAR-2 results in this range will need to be regarded as being potentially alkali-reactive.

These tentative criteria refer to the 'long thin' specimen size presently given in AAR-2, although it is probable that the 'short fat' (or 'short thick') specimen size

(40 × 40 × 160 mm) will become preferred in due course, particularly as this is the recommended specimen size in AAR-5. At present, optional versions of AAR-2 are available for both the long thin (AAR-2.1) and short fat (AAR-2.2) specimens. Based on the findings of the EU 'PARTNER' research programme, on average short fat specimens produce lower values than long thin specimens over the same time period; the mean ratio of expansion of short fat to long thin specimens is in the region of 0.75–0.80. However, this ratio may vary considerably; thus, for many aggregates it would give a misleading result if a fixed ratio is used to extrapolate from one type of specimen to the other.

It follows that, in the case of aggregate combinations producing AAR-2 results (after the standard 16-day test) of 0.10 % or higher for long thin specimens (AAR-2.1) or 0.08 % or higher for short fat specimens (AAR-2.2), precautions will probably need to be taken to minimise the risk of ASR damage to any concrete in which the material is used unless concrete prism testing or field performance indicates otherwise.

There is some evidence, for example from Argentina, Australia [1] and Canada [2], that some slowly reactive aggregates are not detected using the above criteria. In Australia, a limit of 0.10 % after 21 days of storage of long thin specimens in 1 M NaOH solution at 80 °C is used, and is found to correlate with the performance of slowly reactive aggregates in concrete structures; this would equate to 0.08 % at 14 days of storage.

It has been suggested that assessment of the rate of expansion might be an alternative method for interpreting the AAR-2 test, especially in the case of uncertain results, and tentative recommendations for this approach are given in the annex to the AAR-2 method (also see [3], for possible criteria).

7 38 °C Concrete Prism Testing: AAR-3

A 38 °C concrete prism test method for ASR is given in AAR-3. This test can be used in two ways: as a standard test for evaluating the alkali-reactivity of an aggregate combination (AAR-3.1) or as a test for establishing the 'alkali threshold' of a particular aggregate combination (AAR-3.2).

In AAR-3.1, coarse and fine test aggregates are tested together in a standard mix combination and, where pessimum behaviour is suspected (or where it is unknown whether a pessimum behaviour might be expected), repeat tests can be carried out in which the coarse and fine fractions are variously replaced by a non-reactive material. In some cases, it might be considered more desirable to conduct the tests using the actual aggregate combination planned for a particular project, although, in such cases, the usual interpretation criteria could be less applicable.

The test should always be carried out using the cement and alkali contents stipulated in AAR-3.1, including the higher cement content permitted for certain types of aggregate combinations. The interpretation criteria suggested below for

AAR-3 would not be in any way applicable to concrete mixes with lower cement and/or alkali contents.

In AAR-3.2, the test is used as a means of establishing the alkali threshold at which a particular aggregate combination begins to exhibit a deleterious expansion. In this test, four concrete mixes with alkali levels in increments normally between 2 and 5 kg/m^3 Na_2O eq. are tested. The alkali threshold is the alkali level at which there is a deleterious expansion (according to the criteria discussed below). This threshold is used to designate the aggregate reactivity class of the combination according to AAR-7.1 (the international specification for minimising the risk of ASR). It has been suggested that the leaching of alkalis from the relatively small concrete specimens used in this test can result in the alkali threshold determined by this method being higher than that found in field concretes. In using these AAR-3.2 results, therefore, it is recommended that a 'safety margin' is applied to the result to allow for the known differences between laboratory and field specimens, experimental uncertainty (e.g. from alkali leaching) and site batching variability. National specifications making use of such limits will need to decide on an appropriate safety margin from local experience [4].

Criteria for the interpretation of the results of AAR-3 have not yet been finally agreed. However, on the basis of trials carried out by RILEM on aggregate combinations of known field performance from various parts of the world, it seems that results in the test (usually after 12 months) of less than 0.05 % are likely to indicate non-expansive materials, whilst results exceeding 0.10 % indicate expansive materials. It is not currently possible to provide definitive interpretative guidance for results in the intermediate range 0.05–0.10 % and, for all practical purposes in the absence of additional local experience, aggregates yielding AAR-3 results in this range will need to be regarded as being potentially alkali-reactive. Since the RILEM interlaboratory trial, modifications to the method have been made on the basis of further experience and the results of the PARTNER programme. It is believed that these will have the effect of increasing expansions, especially at longer ages, so, if anything, these criteria will be even safer.

It follows that, in the case of aggregate combinations producing AAR-3 results of 0.05 % or higher (after 12 months), in the absence of local experience to the contrary, precautions should be taken to minimise the risk of ASR damage to any concrete in which the material is used. Again, there is some evidence that a lower criterion at 12 months (perhaps 0.04 or even 0.03 %) might be applicable for some slowly reactive aggregates.

In many cases, expansion will have ended or the rate of expansion become greatly reduced by the end of the standard 12-month test period. However, in some cases, expansion might still be occurring at 12 months, suggesting that the above criteria possibly could be exceeded during an extended period of testing. At present it is not possible to provide definitive guidance on the interpretation to be placed on such behaviour in the test, but it is suggested that, if time permits, testing may be continued until expansion ceases or it has become clear whether or not the criteria will be exceeded. If such continued testing is not practicable, a judgement will need to be made, from the inspection of the shape of the expansion curve up to

12 months, as to whether or not the criteria would be likely to be exceeded during further testing. In such a case, it is suggested that, in the absence of any relevant local experience, the result should be deemed to have exceeded the criteria if it is estimated that these criteria would be likely to be exceeded during an extension of testing to 24 months.

8 60 °C Concrete Prism Testing: AAR-4.1

A 60 °C concrete prism test method for ASR has been developed as AAR-4.1 and has been assessed by an international trial as an accelerated version of the AAR-3 test for evaluating the reactivity of an aggregate combination.

Criteria for the interpretation of the results of AAR-4.1 have not yet been finally agreed. However, on the basis of an initial assessment of the AAR-4.1 trials carried out by TC 191-ARP on aggregate combinations of known field performance from various parts of the world, it seems that a maximum expansion in the test of 0.03 % at 15 weeks indicates a non-reactive aggregate combination. It follows that, in the case of aggregate combinations producing AAR-4.1 results greater than 0.03 % at 15 weeks, in the absence of local experience to the contrary, precautions should be taken to minimise the risk of ASR damage to any concrete in which the material is used.

9 Carbonate Aggregate Testing: AAR-5

An accelerated screening test procedure for aggregates comprising or containing carbonate material has been developed as AAR-5 and has been assessed by an international trial. In this procedure, the aggregate material is subjected to testing using both the AAR-2 mortar-bar test and a new derivative test using ‘concrete-bar’ specimens, in which a 4/8 mm aggregate grading is used instead of the 0/4 mm grading used in AAR-2. In this application, both the AAR-2 and AAR-5 procedures employ ‘short fat’ specimens (nb the term ‘concrete-bar’ should not be confused with ‘concrete prism’).

Interpretation of the AAR-5 findings is based upon comparing the results of the two test methods. In typical ASR, the mortar-bar (AAR-2) method may be expected to produce greater expansion than the ‘concrete-bar’ (4/8 mm aggregate) method. However, investigations and trials have shown that expansion is greater in the ‘concrete-bar’ (4/8 mm aggregate) test in the case of carbonate aggregates (i.e., aggregates comprising crushed carbonate rock or natural aggregates containing a substantial proportion of carbonate rocks and minerals) that have been associated with carbonate-related expansion in concrete structures. Additionally, it has been found that these materials are not necessarily identified using the AAR-2 method alone. Therefore, in the AAR-5 procedure, if the ‘concrete-bars’ (4/8 mm aggregate) expand more than the conventional AAR-2 mortar-bars (0/4 mm aggregate), the reactivity of the aggregate is probably not of the normal ASR type and further investigation using the longer-term AAR-4.1 or AAR-3 concrete prism tests will be required.

Interpretation of the comparison between the AAR-2 and AAR-5 results may be summarised as follows (in all cases referring to the 'short-fat' specimen option):

- AAR-2 > 0.08 %:
  - AAR-5 < AAR-2 = potential ASR
  - AAR-5 ≥ AAR-2 = possible combination of ASR & carbonate reaction
- AAR-2 < 0.08 %:
  - AAR-5 ≥ AAR-2 = possible carbonate reaction
  - AAR-5 < AAR-2 = no further testing

Further guidance on the specialised assessment of carbonate rocks and aggregates for reactivity potential is given in Annex A.

10 Precision of the RILEM Methods

The overall precisions of the three expansion RILEM methods were assessed in an inter-laboratory trial as part of the PARTNER Programme [5] following the procedure set out in ISO 5725-94 [6]. Eight laboratories took part in each assessment using three aggregates chosen to give a low, medium and high value in the particular test. The aggregate samples were prepared and distributed by the organizing laboratory. On the basis of this trial the details of the methods were improved so it can be expected that the present RILEM methods will have better precision.

The results of the trial are summarized below:*

Expected reactivity for aggregate in inter-laboratory trial	Low/medium	Medium	High
AAR-2 (long thin bars, 14 days)			
Measured mean expansion *m* (%)	0.133	0.420	0.375
Repeatability COV(sr) (%)	6.8	2.70	1.9
Reproducibility COV(sR) (%)	17.1	22.0	11.0
AAR-3 (using wrapped prisms)			
Measured mean expansion *m* (%)	0.059	0.181	0.260
Repeatability COV(sr) (%)	15.3	13.8	20.7
Reproducibility COV(sR) (%)	49.3	53.7	54.3
AAR-4 (wrapping method)			
Measured mean expansion *m* (%)	0.107	0.130	0.170

(continued)

(continued)

Expected reactivity for aggregate in inter-laboratory trial	Low/medium	Medium	High
Repeatability COV(sr) (%)	11.20	9.1	16.4
Reproducibility COV(sR) (%)	33.1	26.4	24.2

**General definitions:*

r = repeatability. This is a measure to determine the spread in results obtained between the individual prisms, tested at the same laboratory, same aggregate combinations and same concrete.

R = Reproducibility. This is a measure to compare the difference in the mean value obtained between the different laboratories.

COV(sR) = Coefficient of variation for the Reproducibility. By using the coefficient of variation (COV) one relates the spread to the actual expansion. The COV is the standard deviation divided by the mean value. The COV(sR) is thus used to compare the difference in the spread between the laboratories. Similarly, the COV(sr) is used to compare the spread within one single test carried out at one laboratory. In this case, the spread between the three prisms will determine whether it is correct to average their results or if the test has to be remade or only two of results should be averaged; all depending on the instructions in the test method.

10.1 Rilem AAR-2

The precision of this method was found to be quite good. At the proposed limiting value of 0.10 % for the long prisms the band of uncertainty is less than 0.025 % so, in precision terms, this method is well able to differentiate reactive from non-reactive materials.

10.2 Rilem AAR-3

The results concerning this method indicated that the repeatability is good, whereas the reproducibility is quite poor. The precision is, however, good enough to discriminate between non-reactive and reactive materials. The reproducibility is shown to be about half of the expansion value. At the level of 0.05 %, tentatively suggested as the limiting value to differentiate between reactive and non-reactive aggregate combinations when using AAR-3, there is therefore a band of uncertainty of 0.025 %, and the lowest result for a reactive aggregate should exceed 0.075 %. This was found to be the case for all the aggregates which react in normal timescales and where their record of field reactivity is well established. However, the AAR-3 method is not so well suited to identifying "slowly" reactive aggregates unless the time period of the test is extended. It should be noted that in this trial a variation of the method in which the prisms were wrapped was used. This procedure is not now recommended.

10.3 RILEM AAR-4 (Wrapped Prism Method)

Testing according to the AAR-4 method resulted in a better overall precision than AAR-3, and was able to detect smaller differences in reactivity than the AAR-3 method. Like AAR-3, the precision of the AAR-4 Alternative method is sufficient to distinguish between non-reactive and reactive materials, but not between smaller differences of reactivity. At the tentatively proposed limiting value for differentiating reactive and non-reactive aggregates, 0.03 %, the band of uncertainty is less than 0.01 %.

11 Releasable Alkalis

Work is under way in RILEM to develop a standardised test method (AAR-8) for assessing the releasable alkali content of aggregates and this is at the stage of trials of the procedure in various specialist laboratories [7]. Various methods have been suggested and used previously, mostly based upon extraction by a calcium hydroxide solution, but none of these is considered adequately to replicate the possible release of alkalis within concrete. AAR-8 is based upon extraction using an alkali solution saturated with calcium hydroxide and early results suggest that the results may be regarded as meaningful. In due course, RILEM expects to be able to provide guidance on the interpretation and application of AAR-8 findings.

12 Conclusions

Petrographical examination (AAR-1.1 & 1.2) should be carried out in all cases. On some occasions this will lead directly to definitive outcomes, either Class I 'unlikely to be alkali-reactive', or Class III 'very likely to be alkali-reactive'. In many cases, petrographical examination will lead to an indefinite outcome, Class II 'potentially alkali-reactive', and further testing will be required.

Siliceous aggregates (and carbonate aggregates with a significant siliceous content) may be further assessed for ASR, usually using first the short-term (2 or 3 weeks) screening test (AAR-2), then the 60 °C accelerated concrete prism test (AAR-4.1), which can be interpreted after 15 weeks. If required, the longer-term (12 months) 38 °C concrete prism test (AAR-3) may be carried out. The findings of the concrete prism tests should always take precedence. The AAR-2 test cannot be used for Class II aggregates containing porous flint (chert) as a potentially reactive constituent and the criteria for some slowly reactive aggregate types might need to be modified.

Carbonate aggregates (and siliceous aggregates with a significant carbonate content) may be further assessed using the AAR-5 short-term (2 or 3 weeks) screening procedure, which will identify any aggregate reactivity that is probably not of the normal ASR type and indicate when further investigation using the longer-term AAR-4.1 or AAR-3 concrete prism tests will be required. Specific guidance on assessing carbonates is given in Annex A.

In addition to assessment of an aggregate combination for reactivity potential, the aggregates may be tested for releasable alkali content using the AAR-8 method that is being developed.

Annex A: Assessment of Carbonate Rock Aggregates for Reactivity Potential

A1 Scope

This Annex describes procedures for the assessment of potentially reactive carbonate rocks in concrete. The procedures include those in AAR-1.1 & 1.2, which does not include specific guidance on the assessment of carbonate rocks for reactivity in concrete. As a result of undergoing the procedures described in this Annex, carbonate rocks should be classified according to one of the following classes:

- Very unlikely to be alkali-reactive—Class I
- Alkali-reactivity uncertain—Class II
- Very likely to be alkali-reactive—Class III

It is very important that the petrographic analysis is carried out by a qualified geologist with experience of materials used for concrete and good local knowledge of alkali-reactive aggregates, minerals and in this case especially carbonate rocks.

A2 Definitions

A2.1 Alkali Carbonate Reaction (ACR)

Chemical-physical expansive reaction in concrete between certain impure coarse grain-sized dolomitic carbonate rocks and the cement paste. The reaction appears to be associated with dedolomitization and an associated swelling reaction, but is not yet fully understood and documented.

Note 1: *The reaction might occur concurrent with Alkali-Silica Reaction (ASR) caused by the same aggregate particle. Some researchers suggest that ASR is the only expansive reaction in reacted carbonate aggregates.*

A2.2 Carbonate Rock

A rock composed of more than 50 % by mass of carbonate minerals such as calcite or dolomite. They are sedimentary or metamorphic, and very rarely igneous (carbonatites) in origin.

A2.3 Carbonate Sedimentary Rocks

Calcareous rock is a rock containing an appreciable amount of calcium carbonate. It can be sedimentary limestone (e.g. chalk, tufa or calcarenite). Dolomitic limestone contains 10–50 % dolomite and 50–90 % calcite. Dolomite rock (dolomite) contains more than 50 % of the mineral dolomite. Dolomite occurs in crystalline and microcrystalline forms. The term "dolostone" is synonymous with dolomite rock, but has not gained universal acceptance.

Carbonate rock deposits can often contain intermixed layers of clay, shale, sandstone or siltstone. Silicification of carbonate rock deposits with dispersed crypto- and micro-crystalline quartz or opal is not uncommon.

A2.4 Metamorphic Carbonate Rocks

Metamorphic carbonate rock is termed marble. Dolomitic marble is composed mostly of the mineral dolomite.

A2.5 Dedolomitization

A process resulting from chemical weathering, diagenesis or metamorphism, wherein part or all of the magnesium component in a dolomite or dolomitic limestone is consumed in the formation of magnesium hydroxides and silicates (e.g. brucite, forsterite), resulting in an enrichment in the calcite content.

A2.6 Chemical Reaction of Dolomite in Concrete

Dolomite can be unstable in concrete under certain conditions. The instability and decomposition of dolomite in concretes produce more stable phases, such as calcite and brucite. It could be the result of the following reaction:

$$\underset{\textit{dolomite}}{CaMg(CO_3)_2} + 2M(OH) = \underset{\textit{calcite}}{CaCO_3} + \underset{\textit{brucite}}{Mg(OH)_2} + M_2CO_3$$

where M is an alkali metal.

It remains uncertain whether or not this dedolomitization process alone can sometimes cause damage to concrete.

A3 Principles

The AAR-1.1 & 1.2 petrographic examination method describes the visual recognition and quantification techniques for rocks and mineral constituents of aggregates sources with special emphasis on their potential for alkali reactivity. This Annex gives supplementary information and methods for assessing carbonate rocks for potential reactivity in concrete.

Thin-sections (optionally polished thin-sections) stained for carbonate rocks should be prepared and used to determine the types of carbonate rocks. The procedures for carbonate rocks are summarised in Fig. A.1. The procedure is generally used for aggregates originating from quarries dominated by carbonate rocks.

Note 2: *Crystalline carbonate rock without dolomite and impurities should be assessed unlikely to be reactive and further testing is not necessary. Carbonate aggregates intended to be used only as fines (sand) in concrete are unlikely to exhibit ACR but would still need to be assessed for ASR potential.*

The procedure allows for the additional use of 3 optional methods when carbonate rocks have been identified in thin section:

(1) X-ray fluorescence analysis (XRF),
(2) X-ray diffraction analysis (XRD),
(3) Scanning Electron Microscopy (SEM) with Energy Dispersive X-ray analysis (EDX).

Note 3: *Detection of dolomite and the potential degree of dedolomitization (and reaction products) can be made using the following techniques:*

1. *XRD, using an internal standard of very well known d-spacing to determine the d-spacing of dolomite.*
2. *Petrography, using alizarin-red dye, to determine zoning, crystal shape, partial dedolomitization or iron oxides stains as well as EMPA, SEM/EDX.*

Rather than proceeding with XRF analysis, XRD analysis and/or SEM/EDX (or WDX, see A6) analysis, this procedure also includes direct application of screening tests according to AAR-5 or even the longer-term AAR-3 or AAR-4.1 concrete expansion tests.

A4 Assessment Using XRF Analysis (Optional Method)

When carbonate minerals are identified by thin-section, an X-ray fluorescence (XRF) analysis on the bulk sample might be carried out. Minimum elements to be

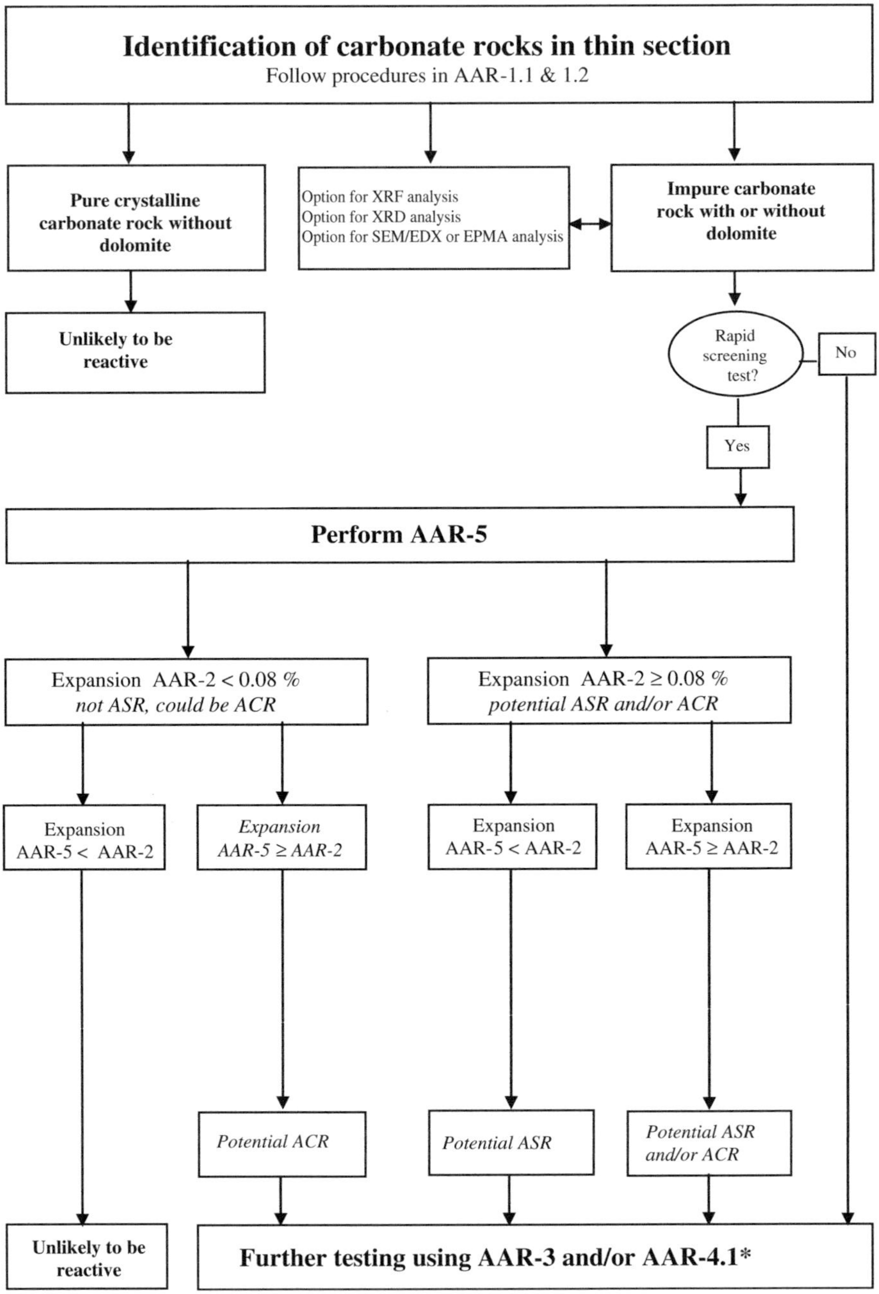

Fig. A.1 Flow chart for AAR assessment of carbonate rocks. * There is limited experience in using AAR 4.1 for carbonate rocks/aggregates

analysed are calcium and magnesium to indicate the carbonate minerals (calcite and dolomite), and aluminium to indicate the possible presence of clay minerals (see also Note 5 below).

For the assessment of reactivity, the calcium/magnesium oxide ratio should be calculated and the ratio plotted in Fig. A.2 against the aluminium oxide content. Two possibilities will result, based upon empirical observations in Canada, namely to be "considered potentially expansive" or "considered non-expansive". Because of limited experience with the method outside Canada, a further assessment of possible expansion should be carried out according to AAR-5.

A5 Assessment Using XRD Analysis (Optional Method)

When carbonate minerals are identified by thin-section (see AAR-1.1 & 1.2), an X-ray diffraction (XRD) analysis on a bulk sample might be carried out. Generally the major (>5 %) and minor minerals (<5 %) are analysed and reported. Normally XRD analysis is used for qualitative identification of crystalline minerals (Note 4), but can also be used for semi-quantitative measurements.

The carbonate minerals magnesite, dolomite, ferroan dolomite and calcitic dolomite are considered indicative of potentially ACR reactive material. At present, the identification of any detectable quantity of these phases should classify the aggregate sample as "potentially expansive" and further testing according to AAR-5 should then be performed (see Fig. A.1).

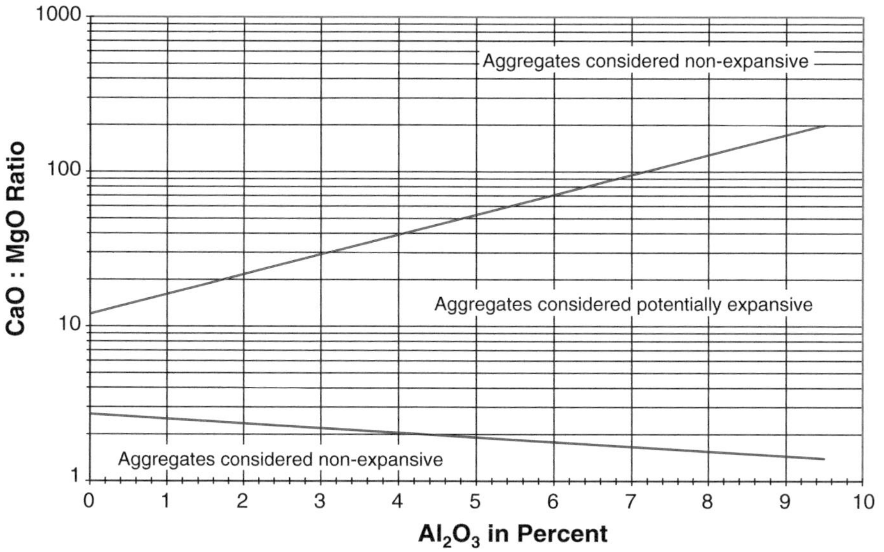

Fig. A.2 Suggested interpretation of XRF analysis findings. Figure modified from C.A. Rogers [8]

If ACR indicative minerals have not been identified by the XRD analysis, ACR is unlikely to occur, but ASR is still possible. In that case testing according to AAR-2 and/or AAR-3 and/or AAR-4.1 could be performed (see Fig. A.1).

Note 4: *XRD analysis, which identifies only crystalline materials, will not be able to characterize amorphous constituents (e.g. opal-A, glass, or other non crystalline constituents).*

Note 5: *In some carbonate rocks, clay minerals can occur that might also cause problems and non-AAR expansion in concrete. When necessary, clay minerals can be characterised using specialised XRD. In sedimentary carbonate rocks, the total alumina content is also a useful indicator of the amount of clay minerals, since alumina is normally only derived from clay minerals in the absence of feldpars (authigenic or detrital); clay mineral content is approximately 3 x the content of alumina (Al_2O_3).*

A6 Assessment Using SEM/EPMA Analysis (Optional Method)

When carbonate minerals are identified by thin-section analysis (see AAR-1.1 & 1.2), examination by use of Scanning Electron Microscopy (SEM) and/or Electron Probe Micro Analyser (EPMA) can be carried out. This examination should be carried out by qualified personnel with knowledge of these techniques. It is recommended to use polished thin-sections or polished samples. Elements can be detected and quantified by use of Energy Dispersive X-ray (EDX) analyses and minerals by use of Wavelength Diffraction X-ray (WDX) analysis.

Interpretation of results obtained by EDX is the same as given in Fig. A.2. Interpretation of results obtained by WDX analysis is the same as given above for XRD.

A7 Assessment of Reactivity According to AAR-5

An accelerated screening test procedure for aggregates comprising or containing carbonate material has been developed as AAR-5 and has been assessed by an international trial.

In this procedure, the aggregate material is subjected to testing using both the RILEM AAR-2 mortar-bar test and a new derivative test using 'concrete-bar' specimens, in which a 4/8 mm aggregate grading is used instead of the 0/4 mm grading used in AAR-2. In this application, both the AAR-2 and AAR-5 procedures employ 'short fat' prism specimens (40 × 40 × 160 mm).

The interpretation of the AAR-5 findings is based upon comparing the results of these two test methods. In typical ASR, the mortar-bar (AAR-2) method may be expected to produce greater expansion than the 'concrete-bar' method. However, investigations and trials have shown that expansion is greater in the 'concrete-bar' test in the case of carbonate aggregates that have been associated with expansion in concrete structures, and also that these materials are not necessarily identified using the AAR-2 method alone. Therefore, in the AAR-5 procedure, if the concrete-bars expand more than the mortar-bars, the reactivity of the aggregate is probably not that of the normal ASR type and further investigation using the longer-term AAR-4.1 and/or AAR-3 concrete prism tests will be required.

Interpretation of the comparison between the AAR-2 and AAR-5 results (both using the 'short-fat' specimen option) may be summarised as follows (where appropriate, the 0.08 % expansion criterion shown may be replaced by a locally determined value):

- RILEM AAR-2 ≥ 0.08 % and:
  - AAR-5 < AAR-2 = potential ASR
  - AAR-5 ≥ AAR-2 = possible combination of ASR and ACR
- RILEM AAR-2 < 0.08 % and:
  - AAR-5 ≥ AAR-2 = possible ACR
  - AAR-5 < AAR-2 = unlikely to be reactive (no further testing necessary)

A8 Assessment of Reactivity Using AAR-3 and/or AAR-4.1

If potential ASR and/or potential ACR are detected, the longer-term (at least 12 months) 38 °C concrete prism test (AAR-3) may be carried out. Concrete test prisms are prepared from the aggregate combination and are stored in warm, humid conditions for 12 months to promote any alkali-silica reaction or alkali-carbonate reaction. The findings of the concrete prism tests should always take precedence over the results of AAR-2 or AAR-5.

Alternatively, aggregates may be assessed for ASR or ACR using the 60 °C concrete prism test AAR-4.1, which can be interpreted after 15 weeks. It is envisaged that the AAR-4.1 method might be used as an accelerated version of the AAR-3 test. However, at present there is only limited experience of using the 60 °C concrete prism method for ACR detection.

A9 Limitations and Need for Research

The suggested test procedures are based on the present knowledge of ACR. The optional XRF method (see Fig. A.2) is based primarily on Canadian experience and some rarer cases internationally. Therefore, experience and testing of carbonate rocks other than Canadian materials are needed to validate or revise the Canadian procedures and criteria.

The mechanism of ACR is not fully understood and more research is needed on this issue. Carbonate rocks are internationally important and widely used aggregate types for concrete. The guidance given in this Annex will hopefully be a step forward in producing durable concrete with carbonate rocks, but will need to be reviewed periodically and updated as appropriate.

Annex B: Guide to Reference Materials

B1 Preamble

This guide is intended to provide assistance to any laboratories undertaking the RILEM expansion tests, using either mortar-bar or concrete-bar specimens (AAR-2 & AAR-5) or concrete prism specimens (AAR-3 & AAR-4.1). It includes information on the use of reference cement or aggregate materials and various accessories required for conducting the tests.

B2 Introduction

The use of reference cement and aggregate materials is not mandatory in the AAR-2, AAR-3, AAR-4.1 and AAR-5 test methods. However, in any testing, the use of reference materials, with known and constant properties or behaviour, may be useful, or stipulated, in certain circumstances, including the following:

- to establish the reliability and accuracy of a new test procedure;
- to assess the competence of a laboratory or the testing personnel;
- to provide reassurance in the case of tests yielding variable results;
- to provide controls for direct comparison with material under evaluation.

In particular relation to the three TC 219-ACS expansion tests for alkali-aggregate reaction, reference materials may be specifically used as follows:

- **Reference High-Alkali Cement**: to minimise any variations arising from using cements of different sources, compositions and properties;
- **Reference Reactive Aggregate**: to provide reassurance to laboratories undertaking tests for the first time, to enable routine checking of testing facilities or their personnel and for use in inter-laboratory precision experiments;
- **Reference Non-Reactive Aggregate**: to enable a baseline movement to be established for testing facilities and for use in programmes for identifying any pessimum behaviour.

B3 Selected Reference Materials

B3.1 High-Alkali Cement

A source of suitable high-alkali Portland cement has been selected, as follows:

- **Norcem, Norway**:
  Cite reference: RILEM reference cement
  Contact: Dr Knut Kjellsen,
  Norcem AS, R&D Department,
  3950 Brevik, Norway
  Telephone: +47 35 57 20 00
  Fax: +47 35 57 04 00
  E-mail: knut.kjellsen@norcem.no
  Minimum quantity: 25 kg (& supplied in multiples of 25 kg)

Property data for this cement are given in Table B.1.

B3.2 Reactive Aggregates—ASR

Many 'reactive' aggregates have been used in experimental research into ASR, variously using natural and synthetic materials. RILEM recommends that a natural aggregate should be selected and that the preferred material should have exhibited a sensibly uniform behaviour in various test methods. After reviewing the options, a crushed siliceous limestone from Spratt's Quarry, near Ottawa in Canada, has been selected.

A stockpile of material from the appropriate strata at Spratt's Quarry has been established by the Ontario Ministry of Transportation, who are prepared to supply modest amounts, as follows:

- **Ontario Ministry of Transportation**:
  Cite: 20–5 mm crushed Spratt's aggregate
  Contact: Mrs Carole Anne MacDonald, Petrographer, Soils and Aggregates Section
  Building C, Room 220, 1201 Wilson Avenue, Downsview, Ontario, M3 M 1J8, Canada
  Telephone: +1 416 235 3738
  Fax: +1 416 235 4101
  E-mail: caroleanne.macdonald@ontario.ca
  Minimum quantity: 25 kg

Table B.1 Property data—reference high-Alkali Cement*

Source	Norcem A/S, Norway
Type	CEM I 42.5 R
Description/sample	Quality declaration
Date (day/month/year)	08/2/2010
CHEMICAL ANALYSIS	% by mass
Loss on ignition	1.1
Insoluble residue	na
Silica, SiO_2	19.8
Alumina, Al_2O_3	5.2
Ferric oxide, Fe_2O_3	3.5
Lime, CaO	61.7
Magnesia, MgO	2.7
Sulfur trioxide, SO_3	3.7
Potash, K_2O	1.16
Soda, Na_2O	0.52
Chloride, Cl	0.05
Phosphorous pentoxide, P_2O_5	na
Chromium, Cr^{6+}	0.00 mg/kg
Free lime	1.4
Total alkali, Na_2O eq.	1.28
Lime saturation factor	na
C_3S	na
C_2S	na
C_3A	na
C_4AF	na
Gypsum	na
Limestone	0.9
PHYSICAL PROPERTIES	EN 196
Fineness, Blaine, m^2/kg	573
Sieve analysis: >90 μm, % by mass	0
Soundness, Le Chatelier, mm	0
Soundness, autoclave, %	na
Setting times, min: initial final	105 na
Compressive strength, MPa: 1 day 2 days 3 days 7 days 28 days	 31 37 na 48 54

*These data are summarised from a certificate supplied to RILEM TC 219-ACS by the manufacturer. Data for presently available batches should be obtained from the manufacturer
na = not available

Geological information, together with some analytical and test data, is given in Figs. B.1, B.2 and Tables B.2, B.3.

A precision trial using an accelerated mortar-bar test [9] was carried out in North America in 1995 (Rogers et al. 1996). This indicated an average 14-day expansion of about 0.42 %, with all compliant laboratories yielding results greater than 0.30 %. A further study with new samples in 2007 produced a similar average 14-day expansion of 0.39 % [10].

In a concrete prism test (CSA method), using cement with an alkali content of 1.25 % (as Na_2O eq.) and 38 °C storage, expansion values with Spratt's coarse aggregate (and non-reactive sand) at 1 year have been reported in the range 0.08–0.16 %. An inter-laboratory concrete prism test study (CSA method), using mixtures of Spratt's coarse aggregate and non-reactive sand, produced average expansion values in the range 0.16–0.18 %, depending upon mix details and storage conditions [11].

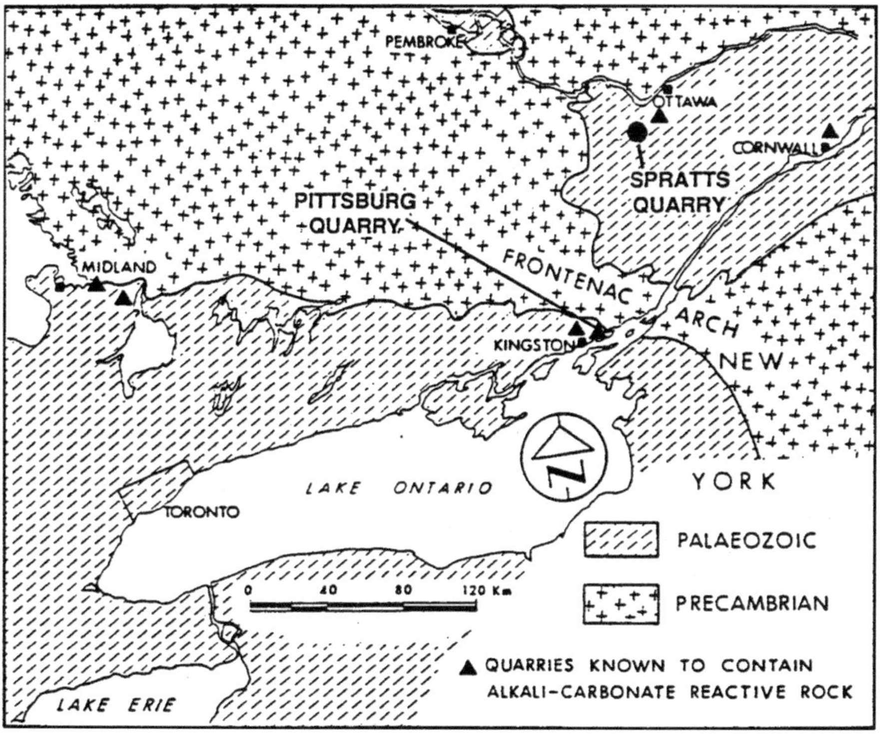

Fig. B.1 Geological map showing location of Spratt's and Pittsburg Quarries. *Reproduced by courtesy of the Ontario Ministry of transportation*

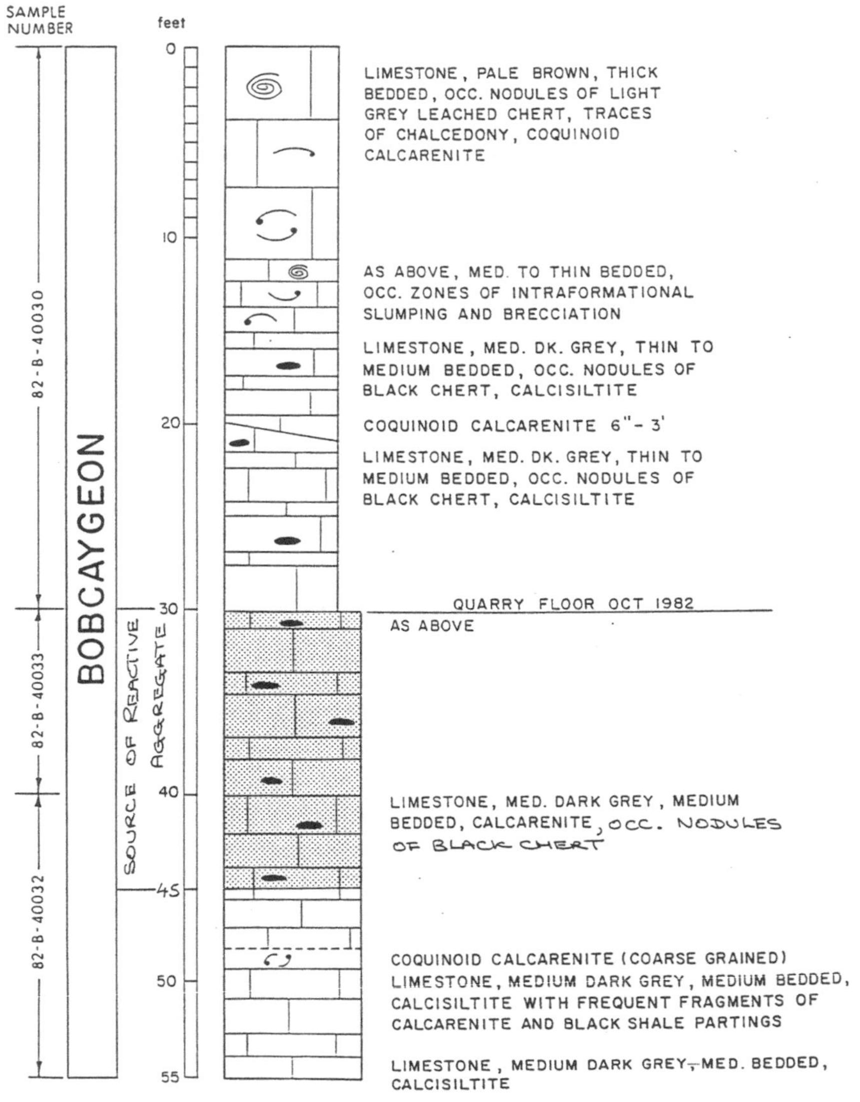

Fig. B.2 Stratigraphic column showing layers exposed in Spratt's Quarry. *Reproduced by courtesy of the Ontario Ministry of transportation*

B3.3 Reactive Aggregates—Carbonate

A stockpile of reactive carbonate aggregate material from the Pittsburg Quarry at Kingston, Ontario, Canada, has been established by the Ontario Ministry of Transportation, who are prepared to supply modest amounts, as follows:

- **Ontario Ministry of Transportation**:
  Cite: 20–5 mm crushed Pittsburg Quarry aggregate
  Contact: Mrs Carole Anne MacDonald, Petrographer, Soils and Aggregates Section
  Building C, Room 220, 1201 Wilson Avenue, Downsview, Ontario, M3 M 1J8, Canada
  Telephone: +1 416 235 3738
  Fax: +1 416 235 4101
  E-mail: caroleanne.macdonald@ontario.ca
  Minimum quantity: 25 kg

The geological location of Pittsburg Quarry is shown in Fig. B.1 and some preliminary analytical and test data are given in Table B.4.

Table B.2 Information and data—reference reactive Spratt's aggregate*

CHEMICAL ANALYSIS[1]	Whole rock	Acid insoluble portion
	% by mass	
Acid insoluble residue	10	100
Silica, SiO_2	8.70	86.92
Alumina, Al_2O_3	0.59	4.24
Titania, TiO_2	0.04	0.21
Phosphate, P_2O_5	0.29	0.45
Ferric oxide, Fe_2O_3	0.58	1.28
Lime, CaO	48.47	0.26
Magnesia, MgO	1.67	0.78
Soda, Na_2O	0.04	0.08
Potash, K_2O	0.08	0.78
Sulfur, S	0.13	1.16
Loss @ 1050 °C	39.55	4.02
Total	100.14	100.18
MINERALOGY[2]	Whole rock	Acid insoluble portion
	phases detected and order of concentration	
Calcite	major	–
Quartz	minor	major
Dolomite	minor	–
Pyrite	nd	minor
Illite (clay mineral)	nd	minor

*These summary data are collated from detailed information held on file by RILEM TC 219-ACS
nd = not detected (below lower level of detection for method)
[1]X-ray fluorescence, by Hung Chen, Canada Cement Lafarge Ltd, Montreal
[2]X-ray diffraction, by Hung Chen, Canada Cement Lafarge Ltd, Montreal

Table B.3 ASR test data—reference reactive Spratt's aggregate

ASTM C289 CHEMICAL METHOD[1]	R_c/S_c millimoles/litre (classification)
300–150 μm (acid insoluble component)	36/307 (deleterious)
<150 μm (acid insoluble component)	52/391 (deleterious)
ASTM C227 MORTAR-BAR TEST[2]	% expansion, range (various storage types)
13 weeks (3 months)	<0.05–0.14
26 weeks (6 months)	<0.10–0.28
39 weeks (9 months)	<0.10–0.34
ACCELERATED MORTAR BAR TESTS[3]	
ASTM C1260 (number of labs = 23)	% expansion, range (mean), standard deviation
After immersion for 14 days	0.276–0.457 (0.389), 0.044
After immersion for 21 days	0.407–0.580 (0.495), 0.058
After immersion for 28 days	0.488–0.700 (0.584), 0.058
CSA A23.2-25 (2009) (number of labs = 28)	
After immersion for 14 days	0.307–0.486 (0.372), 0.042
After immersion for 21 days	0.383–0.592 (0.482), 0.052
After immersion for 28 days	0.450–0.725 (0.582), 0.068
RILEM AAR-2 (number of labs = 3)	
After immersion for 14 days	0.23–0.369 (0.291),[4]
After immersion for 21 days	0.357–0.490 (0.410),[4]
After immersion for 28 days	0.500–0.621 (0.547),[4]
CONCRETE PRISM EXPANSION TESTS[3]	
CSA A23.2-14A and ASTM C 1293 (number of labs = 35 and 36)	% expansion, range (mean), standard deviation
Using control sand, 12 month expansion at 38 °C	0.128–0.334 (0.204), 0.050
Using local sand, 12 month expansion at 38 °C	0.104–0.268 (0.175), 0.039
RILEM AAR-3 (number of labs = 3)	
Using control sand, 12 month expansion at 38 °C	0.184–0.344 (0.274),[4]
Using local sand, 12 month expansion at 38 °C	0.210–0.308 (0.275),[4]
CSA A23.2-14A and ASTM C 1293 but tested at 60 °C (number of labs shown below)	
Using control sand, 13 week expansion (n = 19)	0.056–0.217 (0.146), 0.041
Using control sand, 26 week expansion (n = 21)	0.075–0.231 (0.165), 0.042
Using local sand, 13 week expansion (n = 22)	0.053–0.204 (0.113), 0.036
Using local sand, 26 week expansion (n = 23)	0.064–0.230 (0.127), 0.041

[1]Grattan-Bellew, P E, July 1987 (whole rock testing gives 128/32, in the innocuous field)
[2]Cement total alkali content 1.17 % as Na_2O eq., Ontario Hydro-MTC study
[3]These data correspond to the results of an interlaboratory study that was carried out in 2007–2009. The study was organized by Chris Rogers (former MTO and now retired/consultant), Carole-Anne MacDonald (MTO) and Benoit Fournier (Laval University)
[4]Number of laboratories was too few to enable a standard deviation to be calculated

Table B.4 Analytical and test data—reference reactive Pittsburg carbonate aggregate

ASTM C586 ROCK CYLINDER TEST[1]	% Expansion
1 day	0.04
3 days	0.08
7 days	0.28
15 days	0.81
28 days	1.72
64 days	3.50
CSA CHEMICAL ANALYSIS	% by mass[2]
CaO	40.9, 42.6, 43.2
MgO	6.29, 6.17, 8.31
Al_2O_3	2.70, 2.08, 1.81
Classification by CaO/MgO Ratio vs. Al_2O_3	Potentially expansive
PETROGRAPHY	Observations @ NRC & CANMET
Texture	Rhombic crystals of dolomite (20–50 μm) in a matrix of micrite and clay minerals
Study by XRD of effect of NaOH treatment	Formation of brucite (after 14 days) & progressive reductions in dolomite & quartz
CSA CONCRETE PRISM EXPANSION TEST[3,4]	% expansion, range (mean %), standard deviation
One year stored at 23 °C in moist room	0.105–0.210 (0.153), 0.0294
Two years stored at 23 °C in moist room	0.158–0.250 (0.193), 0.0350
One year stored at 23 °C in moist room sealed in plastic bag	0.153–0.298 (0.235), 0.0404
Two years stored at 23 °C in moist room sealed in plastic bag	0.211–0.389 (0.295), 0.0476
One year stored in 5 % NaCl solution at 23 °C	0.116–0.292 (0.236), 0.0521
Two years stored in 5 % NaCl solution at 23 °C	0.199–0.399 (0.340), 0.0615
One year stored at 38 °C in sealed box with moisture	0.218–0.466 (0.307), 0.0716
Two years stored at 38 °C in sealed box with moisture	0.235–0.537 (0.375), 0.0862

[1]Good samples are not available to conduct this test. Data will be found on expansion of rock cylinders from this quarry in Williams and Rogers [12] and Rogers [13]

[2]Second set of chemical data for Pittsburg aggregate is unpublished 1993 inter-laboratory (12 laboratories) data from the Ontario Ministry of Transportation, whilst the third set of data is from Shehata et al. [14]

[3]Source of data for concrete prism expansion testing is results of an unpublished report on work conducted in 1991–1992 from Chris Rogers (2011)

[4]CSA A23.2-14A 1986, with 310 kg/m^3 high-alkali Portland cement boosted to 1.25 % Na_2O eq. Number of labs 15–17

B3.4 Non-reactive Aggregate

A suitable non-reactive aggregate is defined using an unusually demanding criterion of less than 0.05 % expansion in the AAR-2 accelerated mortar-bar test.

In the RILEM trials of the AAR-4.1 60 °C concrete prism test, a crushed limestone from Boulonnais in France has been identified for use as the non-reactive reference coarse and fine aggregates. Arrangements have been made for ex-Boulonnais coarse and/or fine aggregates to be stocked and supplied by the following organisation:

- **Carrières du Boulonnais** (http://www.lesgranulatsdugroupecb.com):
  Cite: Coarse and/or Fine Boulonnais Aggregate
  Contact: Ms Sophie Citerne (Export/Industry Sales Manager)
  530 BD du Parc d'affaires Eurotunnel
  62231 Coquelles
  France
  Mobile/Cell: +33 (0)6.08.33.57.76
  E-mail: sciterne@groupecb.com

B4 Test Accessories

B4.1 Storage Containers for Concrete Prisms

The AAR-3 concrete prism test involves the storage of specimens in a suitable container, as defined in the method. One source of suitable containers is as follows:

- **LINPAC Ropak**
  Cite: 5 gallon or 19 litre round container 10540 Talbert Avenue, Suite 200
  Fountain Valley, CA 92708, USA
  Tel: +1 (714) 845 2845
  Web: www.linpac.com
  E-mail: info@linpacpackaging.com

B4.2 Reactor Storage for Concrete Prisms

The recommended storage for concrete prisms in the AAR-4.1 test utilises the reactor system, which can also be used for the AAR-3 test. Information on this apparatus may be obtained from the following:

- **Chaudronnerie Mecanique Generale**:
  27 rue de la Constellation
  Parc St Christophe, BP 8262
  95801 Cergy Ontoise Cedex, France

- **Espo-Sud**:
  Quartier les Ramières, BP37
  07350 Cruas
  France

- **Schleibinger Geräte Teubert u Greim GmbH**:
  Gewerbestr. 4
  D-84428 Buchbach
  Germany
  Tel: +49 8086 94010
  E-mail: schlei@schleibinger.com

- **Ratio TEC Prüfsysteme GmbH**:
  In der Au 17
  88515 Langenenslingen
  Germany
  Tel: +49 7376 9622-0
  E-mail: ratio.tec@t-online.de

References

1. Shayan, A, 2007, Field evidence for inability of ASTM C1260 limits to detect slowly reactive Australian aggregates, *Australian Journal of Civil Engineering*, 3 (1), 13–26.
2. CSA, 2009, Standard practice to identify degree of alkali-reactivity of aggregates and to identify measures to avoid deleterious expansion in concrete, CSA A23.2–27A, *Canadian Standards Association*, Mississauga, Ontario, Canada.
3. Berra, M, Mangialardi, T, Paolini, A E, 2005, Alkali-silica reactivity criteria for concrete aggregates, *Materials & Structures*, 38 (April), 373–380.
4. BCA, 1999, Testing protocol for greywacke aggregates. *British Cement Association*, Document 45.044, Crowthorne, United Kingdom.
5. Lindgard, J, Nixon, P J, Borchers, I, Schouenborg, B, Wigum, B J, Haugen, M, Akesson, U, 2010, The EU *Partner* project – European standard tests to prevent alkali reactions in aggregates, *Cement & Concrete Research*, 40, 611–635.
6. ISO 5725-94; Precision of test methods – determination of repeatability and reproducibility for standard test method by inter-laboratory tests (1994–1998).
7. RILEM Recommended Test Method AAR-8: Determination of Alkalis Releasable by Aggregates in Concrete (in preparation by RILEM TC 219-ACS).
8. Rogers, C A, 1986, Evaluation of the potential for expansion and cracking of concrete caused by the alkali-carbonate reaction, *Cement, Concrete & Aggregates*, CCAGDP, 8 (1), 13–23.
9. ASTM C1260, 1994, Standard test method for potential alkali reactivity of aggregates (mortar-bar method), *American Society for Testing and Materials*, Philadelphia, USA.
10. Rogers, C A, 2007, Progress report on interlaboratory study of Spratt Aggregate #3—December 2007. Soils and Aggregates Section, *Ontario Ministry of Transportation*, Downsview, Canada.

11. Fournier, B, Malhotra, V M, 1996, Inter-laboratory study on the CSA A23.2-14A concrete prism test for alkali-silica reactivity in concrete, In: Shayan, A (Ed.), *Proceedings of the 10th International Conference on Alkali-Aggregate Reaction in Concrete*, Melbourne, Australia, 302–309.
12. Williams, D A & Rogers, C A, 1991, Field trip guide to alkali-carbonate reactions in Kingston, Ontario, *Ontario Ministry of Transportation*, MI Report 145, pp 26.
13. Rogers, C A, 1986, Evaluation of the potential for expansion and cracking of concrete caused by the alkali-carbonate reaction, *Ontario Ministry of Transportation*, Report EM-75, pp 42.
14. Shehata et al., 2011, The effects of supplementary cementing materials on expansion of concrete containing carbonate aggregates of marginal alkali reactivity, *2nd International Engineering Mechanics and Materials Speciality Conference*, Ottawa, Ontario June, 2011.

RILEM Recommended Test Method: AAR-1.1—Detection of Potential Alkali-Reactivity—Part 1: Petrographic Examination Method

1 Foreword

Petrographic analysis should always be the first step in the assessment of the potential alkali-reactivity of concrete aggregates as stated in RILEM AAR-0 (Outline guide to the use of RILEM methods in assessments of aggregates for potential alkali-reactivity).

This RILEM method specifies a general procedure for the petrographic examination of concrete aggregates, to identify rock types and minerals that might react with hydroxyl ions from the concrete pore solution. The method, including sampling, is intended to be used for the routine examination of natural aggregates including sand, coarse gravel, all-in and crushed rock aggregates, in testing laboratories. The method can also be used to quantify the amounts of various mineral and rock types where required. The main objective of the RILEM petrographic method is to determine a classification of a particular aggregate in terms of alkali-reactivity. As a result of undergoing the RILEM petrographic examination an aggregate should be classified as one of the following:

Class I—Very unlikely to be alkali-reactive
Class II—Alkali-reactivity uncertain
Class III—Very likely to be alkali-reactive

Acceptance and experience with reactive constituents differ between countries, and thus, final assessment of reactive constituents should where possible follow national or regional experiences, recommendations and specifications. RILEM AAR-0 also gives guidance on the further testing that may be undertaken on the basis of the findings of the AAR-1.1 petrographical examination.

Petrographers carrying out the method can find assistance in the RILEM Petrographic Atlas (AAR-1.2) [1] which contains micrographs showing the main types of reactive rocks found worldwide. In due course, this method will also be

P.J. Nixon and I. Sims (eds.), *RILEM Recommendations for the Prevention of Damage by Alkali-Aggregate Reactions in New Concrete Structures*,
RILEM State-of-the-Art Reports 17, DOI 10.1007/978-94-017-7252-5_3

supplemented by more detailed information given in a RILEM guide to the use of this procedure.

Note 1: *The main focus of this AAR-1.1 guidance is on the correct identification and—where necessary—modal quantification of alkali-silica reactive (ASR) constituents in aggregate for use in concrete. Specific guidance on petrography and other techniques for assessing carbonates, including potentially alkali-carbonate reactive (ACR) rocks or aggregates, is given in Annex A of AAR-0.*

2 Scope

The RILEM petrographic method describes techniques and instrumentation for the identification of mineral constituents and rock types of aggregate for use in concrete, primarily to determine their potential for deleterious Alkali Silica Reaction (ASR). The protocol to be followed consists of a cascaded sequence of analytical procedures as outlined in the flow chart given in Fig. 1.

The method and the techniques described demand a competent petrographer capable of (1) identifying the constituent minerals and rock types in a sample material, and (2) correctly interpreting these in terms of their potential to develop deleterious ASR, using macroscopic determination, and/or thin-section petrography.

The core of the method comprises two techniques that work on complementary scales of observation, notably:

1. Macroscopic petrography, to identify rock types present and, where required, to determine the modal contents of rock types in coarse aggregates of particle size >4 mm in weight percent by visual inspection, manual particle separation and weighing of the obtained fractions, and
2. Thin-section petrography, using an optical microscope, to determine mineral content and rock types in volume percent by point counting where required. This technique can be applied to fine aggregate particles (<4 mm), fine-grained rock types, as well as on coarse aggregate particles that cannot be unequivocally identified through macroscopic determination.

Together, both techniques should provide a reliable account of the minerals and rock types present in the material concerned. For certain rock types, e.g. those containing finely disseminated reactive minerals, supplementary analytical techniques may be required for reliable determination.

Note 2: *The distinction between coarse and fine in aggregate size varies across different standards, and is hence arbitrary. Here, we assume aggregate >4 mm as coarse (and aggregate ≤4 mm as fine). A division at 5 mm is equally acceptable in this method.*

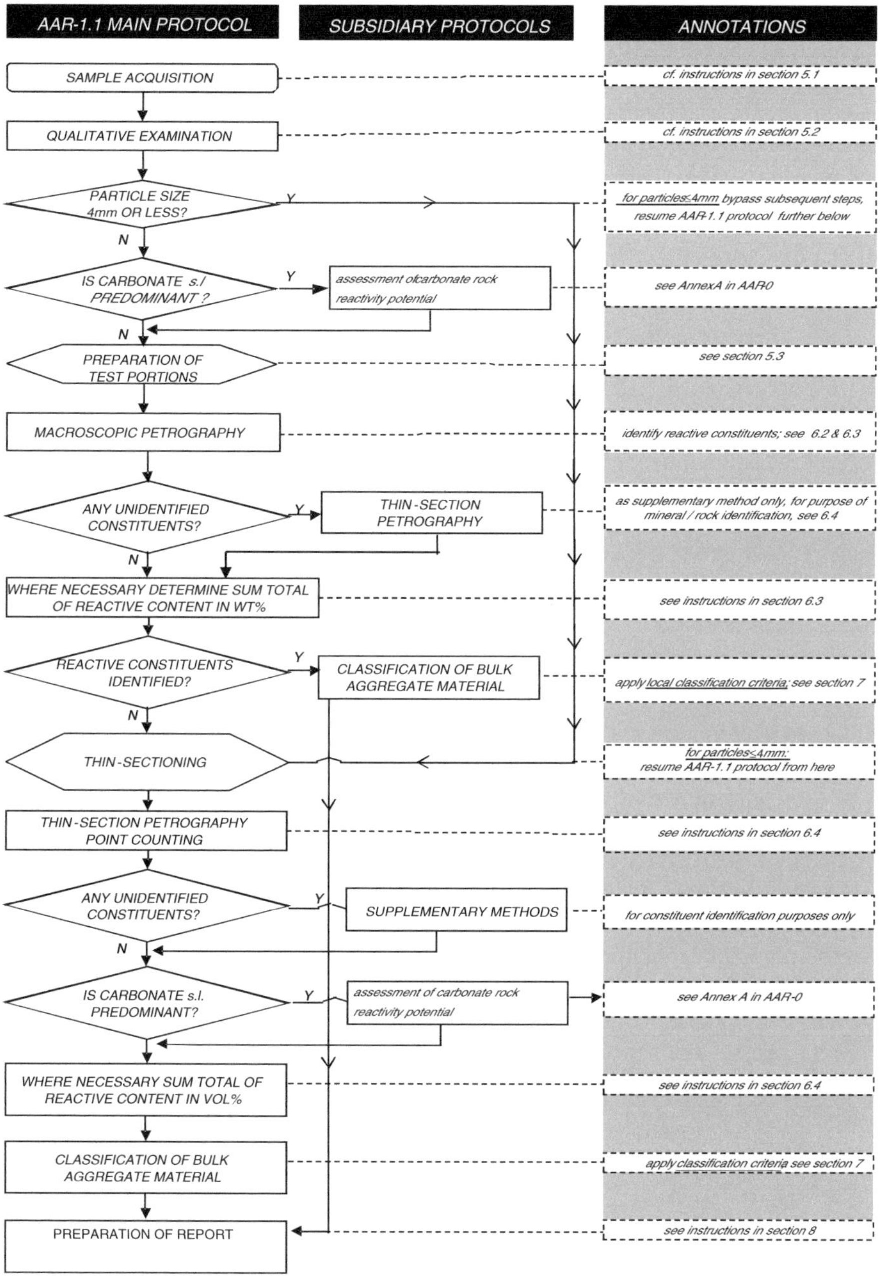

Fig. 1 Flow chart for petrographic assessment of aggregate materials to determine AAR potential

Note 3: *If macroscopic determination alone is able to provide a reliable result, then subsequent assessment by thin-section petrography is not necessary. This requires that all minerals and rock types in the material concerned can be determined unequivocally, and that the material does not contain any constituents of uncertain identity. In addition, if the examination is required to assess compliance with any specified maximum limits on the contents of potentially AAR-deleterious constituents, those contents, including analytical error, must be either well below the limit (i.e., the material will always classify as Class I), or well over the limit (i.e., the material will always classify as Class III).*

3 Preparatory Information

Subject to availability, information on known potentially reactive aggregates may be usefully obtained prior to assessing the sample.

3.1 Geological Information

This information may be in the form of one or more of the following:

- Geological map (e.g. 1:50000 scale) over the deposit area locating the extraction site and indicating main lithological units, location of joints, faults and/or thrust planes, glacial and/or fluvial transport directions, and the like;
- Mapping and geo-research from the published literature;
- Detailed petrographic descriptions of lithological units, including mineral (modal) content and rock names cf. standardized nomenclature (see Sect. 3.3 and Annex A);
- Borehole data containing the major lithological units;
- Quarry map (e.g. 1:1000 scale), plan and time line, showing various lifts and benches extracted over time;
- Influence of weather and climate exposure on the extracted lithologies.

Where practicable, the petrographer should visit the site of extraction and advise on the sampling locations.

3.2 Reactive Aggregates

Any available information on previous alkali-reactivity assessments of the quarry or rocks/lithological units should be made available to the petrographer, including materials such as recommendations, specifications, petrographic descriptions of reactive aggregates, field experiences from concrete structures made with this particular aggregate (methods for detailed assessment of deleterious aggregate in existing concrete are specified in RILEM AAR-6.1, [2]), laboratory results and any other information of importance for the assessment of reactivity.

3.3 Mineral and Rock Nomenclature

Rock types must be named according to internationally acknowledged nomenclature for the systematic naming of rocks, based on mineral modal content in volume percent as well as structural and textural/fabric properties.

Note 4: *Further details on mineral/rock classification and relevant terms and definitions as applied in this AAR-1.1 method are given in Annex 1 and assistance in recognizing and naming the main reactive rock types and minerals is given in the Petrographic Atlas, AAR-1.2* [1].

4 Apparatus

4.1 Preparation of Samples

A rotating divider or riffle box shall be used to reduce the bulk sample material to subsamples of suitable size for further analysis ('test portions'). This enables the production of a representative subsample, compliant with e.g. EN 932-1 [3] Test for general properties of aggregates—Part 1: methods for sampling, or equivalent.

Note 5: *Pile quartering and splitting very easily introduces undesirable artifacts affecting subsample representativeness, and is not recommended.*

Sieves to extract separate size fractions shall comply with ISO 565 [4], ISO 3310-1 [5], ISO 3310-2 [6], ASTM E11 [7], EN 933-2 [8] or equivalent.

Scales shall be capable of weighing the test portions to an accuracy of 0.1 %.

4.2 Macroscopic Assessment

For macroscopic determination of minerals and/or rock types in the sample material, the following tools or instruments may be used:

(a) Triplet or doublet hand lens with 8–12 × magnification;
(b) Standard set of minerals to determine Mohs' scratch hardness;
(c) A pen knife or similar steel probe, for powdering small amounts of material;

(d) A streak plate to determine streak colour;
(e) A magnet or compass, to determine magnetic properties;
(f) Stereomicroscope with magnification 10–80 ×, zoom or incremental;
(g) Drop bottle with a pipette tip containing 10 % dilute hydrochloric acid solution, to test for the presence of calcite that makes the solution effervesce vividly. Dolomite only effervesces with the same solution when finely powdered (e.g., by scraping the surface with a pocket knife) before applying the acid.

4.3 Thin-Section Petrography Using a Microscope

For petrographic assessment of thin-sections using an optical microscope, the following equipment and instrumentation is required:

(a) Laboratory equipment to reduce (crush) particle size to ≤4 mm, if required. The equipment must be cleaned thoroughly before each new sample, to minimize cross-contamination.
(b) Apparatus and machinery for the preparation of (fluorescence-) impregnated thin-sections for petrographic assessment.
(c) A petrographic optical microscope with total magnification ranging from ~20–500×. The instrument shall be equipped with the required accessories for determination of minerals by their optical properties, e.g. a flip condenser, polarizer, rotating stage, gypsum plate, analyzer, Bertrand lens. For fluorescence, the microscope must be equipped additionally with a set of filters (as specified on p.16 of section A3 of Appendix A to CUR-Recommendation 89 (2008) [9]).
(d) A point counting apparatus for use with the petrographic microscope.

5 Sampling and Laboratory Preparation

5.1 Sampling

Sampling should be carried out in accordance with procedures described in EN 932-1 or equivalent other standards such as ASTM D75 [10]. The aggregate sample should be taken from processed material (see Note 6) and must be representative of the bulk material to be used in the concrete. Minimum sizes in kg for bulk samples of natural or processed particulate material are suggested below in Table 1. These are sufficient for qualitative examination or for quantitative examination when the constituent of interest is 20 % or greater and a relative error of ± 10 % is acceptable. If, however, the constituent of interest is present in smaller amounts or a more precise determination is needed, then larger bulk samples will be needed to allow preparation of test samples complying with Table 2.

Record the size of the bulk sample as received in kg, as well as the identification of the sampler (e.g. by petrographer, producer, supplier), and date of sampling. Provide reference to a Certificate of Sampling, as applicable.

Note 6: *Where processed materials are not readily available (e.g., from deposits under development, or currently non-producing quarries), samples shall be acquired from each lithological unit that is expected to be present in the final processed material. These samples shall have an appropriate minimum sample size related to rock grain size (see e.g.* [11]). *The sampling must be carried out by a competent geologist familiar with the lithology of the deposit, consistently following rigorous sampling procedures to ensure that samples are representative.*

Note 7: *Smaller samples than those given in Table* 1 *may be sent by an aggregate supplier to the laboratory for petrographical analysis provided they are sufficient to comply with Table* 2 *and the supplier has sampled a sufficiently large sample according to EN 932-1 or equivalent and has reduced the laboratory sample according to EN 932-2* [12] *or equivalent.*

5.2 *Qualitative Examination of Laboratory Samples*

Initially examine the laboratory sample to ascertain the general characteristics of the aggregate. Distinguish between natural sand and gravel versus crushed rock, and record particle shape (e.g. rounded vs. angular, elongate, flaky). Where aggregates have been produced in part by crushing of a natural gravel (or even boulders), or where aggregates, especially fine aggregates, have been produced by blending natural and manufactured rock materials, state explicitly that the aggregate material is "blended". The presence of potentially deleterious constituents and/or contaminants other than alkali-reactive silica can also be recorded, including micas (e.g., biotite, muscovite, phlogopite), any sulfides, organic matter (e.g., coal, lignite, wood), fine material (e.g., silt, clay, blast residue), particle coatings (e.g., sparingly soluble salts on sea-dredged material), iron hardpan/caliche, encrustations, etc.

Table 1 Minimum sample size of processed particulate material for dispatch to laboratory

Maximum particle size (mm)	Minimum sample mass (kg)
50	200
40	100
20	25
10	10
4	5

Table 2 Minimum representative sample size for laboratory examination

Constituent of interest (%)	Maximum particle size				
	50 mm	40 mm	20 mm	10 mm	4 mm
	Test sample (kg)				Test sample (g)
20	100	45	6	1	50
10	225	95	14	2	115
5	475	200	30	4	250

5.3 *Preparation of Sample Material for Testing*

Reduce the laboratory sample to a suitable test portion or subsample using a riffle box or rotating sample divider, and subsequent sieving. The order of splitting and sieving depends on the mass of the sample and the size fractions to be examined. It is essential to ensure that the material remains representative of the laboratory sample after division into subsamples and test portions (see [11]).

Note 8: *Sample materials may be washed prior to splitting and sieving, using normal tap water over a 0.125 mm sieve, to remove adhering dust and/or contaminants that might interfere with preparation and assessment. A representative sample of the <0.125 mm fraction should be kept for further analysis.*

The minimum required size (in kg) of the representative sample depends on (i) the analytical error deemed acceptable, (ii) the abundance of the alkali-reactive lithology(-ies), which then together determine (iii) the total number of particles to be counted. Representative sample sizes for typical situations are suggested in Table 2 (derived from BS 812-104 [13]).

Note 9: *Samples of all-in aggregate should be divided into their coarse and fine fractions using sieves in a coarse fraction >4 mm and a fine fraction ≤4 mm, maintaining representativeness in terms of number of particles to be counted.*

Note 10: *Sieves must be cleaned meticulously to remove any relics from previous processing, to prevent contamination. Use of an ultrasonic bath and/or a needle to punch out particles stuck in the mesh may be needed.*

5.4 *Preparation of Thin-Sections from Particulate Materials*

Detailed instructions for preparation of a thin-section from particulate materials are given in Appendices A and C of CUR-Recommendation 89 (2008) [9].

6 Petrographic Analysis

6.1 Overall Considerations

Prerequisites

For reliable application of the RILEM petrographic method, the following prerequisites must be fulfilled:

(1) the assessment including macroscopic determination and thin-section petrography is carried out by a competent petrographer familiar with the identification of alkali- reactive minerals and rock types, and able to comprehend and compose a petrographic description of such materials (e.g., see Sect. 3 of ASTM C295-12 [14]);
(2) alkali-reactive constituents present can be identified by macroscopic determination or thin-section petrography using an optical microscope as required;
(3) the sample is representative as received. Any further precautions to maintain a representative sample during preparation and assessment are pointless if the sample as received is not representative.

Amorphous or cryptocrystalline alkali-reactive constituents

In certain rock types (e.g., rhyolite, siltstone, limestone, hornfels, chert/flint), the fine grained reactive silica may be cryptocrystalline or even amorphous beyond the resolving power of an optical microscope. For such rock types, thin-section petrography alone may not be suitable for qualitative nor quantitative assessment of the reactive constituents and must be supplemented with additional analytical methods.

6.2 General Method

The objective of the RILEM petrographic method is to classify the alkali-reactivity of the bulk aggregate (see Sect. 7).

In many cases this can be done by a qualitative or semi-quantitative examination of the aggregate, making use of local experience. Guidance on recognition of reactive rocks is given in the Petrographic Atlas, AAR-1.2 (Note 12) and guidance on acquiring information on reactivity of aggregates from field concrete structures is given in RILEM AAR-6.1 [2]. If, however, there are local (national or regional) recommendations or specifications that give quantitative limits for certain reactive components, then a quantitative assessment, determining the modal contents of the reactive components will be necessary.

The RILEM petrographic method comprises two techniques:

(1) macroscopic petrography as explained in Sect. 6.3, <u>and</u>
(2) thin-section petrography as explained in Sect. 6.4.

6.3 Procedure for Macroscopic Petrography

Macroscopic petrography by particle separation

Macroscopic petrography should be carried out on the coarse fraction >4 mm of all aggregate types, including natural, manufactured and crushed materials, and the coarse fraction of all-in aggregate. Individual particles in the bulk sample are sorted by hand into separate lithological groups based upon macroscopic features and appearance, including e.g., colour, mineral content, particle structure, texture/fabric, morphology, size, density, locally recognized alkali-reactivity, or any other distinguishing property. Where required, the coarse aggregate can be examined in separate size fractions.

Note 11: *Polymictic aggregate materials typically show fractionation (i.e., tendency of various mineral and rock constituents to concentrate into different size fractions) of minerals and rock types among size fractions, due to differences in mechanical (e.g., hardness) or physical (e.g., density) properties, or particle morphology. To determine whether or not fractionation is present, the sample can be (temporarily) subdivided in different size fractions by additional sieving. If any mineral or rock type is predominant in any given size fraction, then fractionation is confirmed.*

Note 12: *Any particles or lithological groups that cannot be identified unambiguously by macroscopic petrography alone should be further assessed by thin-section petrography and/or supplementary techniques.*

Where necessary, the total weight of all particles in each lithological group is recorded and its proportion relative to the bulk sample calculated as weight percent (wt%).

6.4 Procedure for Thin-Section Petrography

Thin-section petrography should be employed for materials where separation into lithological groups by handpicking is not practically feasible or is unreliable, as well as to assess the identity of coarse constituents >4 mm that cannot be determined unequivocally by macroscopic petrography. For quantification of modal contents of constituent minerals and rock types, thin-sections should be assessed by point counting. Petrographic assessment of coarse particles for lithological/mineralogical identification does not require processing, other than trimming to fit the thin-section carrier glass. Where appropriate, larger thin-sections can be useful for examination of coarser particles.

For the reliable quantification of fine aggregate fractions it is recommended that the fine aggregate is split into different size fractions. If fractionation is or might be present then a split into three fractions is preferred; 2–4 mm, 1–2 mm and 0.063–1 mm. Alternatively, where considered appropriate, two fractions can be used, i.e. 2–4 mm and 0.063–2 mm. Exceptionally, a larger number of size fractions can be examined.

Note 13: *The photo atlas in AAR-1.2* [1] *is an essential supplement to the text of this AAR-1.1 procedure. The atlas shows photomicrographs of selected alkali-reactive rock types from around the world, to assist in correct identification. As rock types originally designated as non-reactive are still being identified as alkali-reactive, the photo atlas will consequently not list all alkali-reactive rock types.*

Point counting and reliability

Point counting is a widely accepted method in petrography for quantification of minerals or specific features in thin-sections, and is applied here to assess the relative volume of potentially alkali-reactive constituents.

The thin-section to be assessed by point counting is mounted on to the stage, and moves under the microscope in two perpendicular directions, thus defining a virtual orthogonal grid covering the entire specimen. Grid spacing is to reflect both particle size and packing density as in the specimen. During point counting, only particles located under the cross hairs are counted including edges, both upon entry or exit. Other particles in view but not under the cross hairs are not counted.

Any points falling on mounting resin in between particles are invalid and not counted but skipped: the section is moved forward to the next valid grid point on the edge of or well within a particle. However, any points falling on resin in a vug or void intrinsically belonging to the natural internal porosity of a given particle (i.e., excluding pull-outs or other preparation artifacts) are counted as the hosting lithology.

The minimum number of particles to be counted is dependent on the relative abundance of the (mineral, rock) species of interest in the bulk sample material and the acceptable analytical error as illustrated in Table 3 and explained in Annex B. As an example, at 20 % abundance in the 2–4 mm fraction, if 1000 counts are made the relative error would be 16 %, i.e. ±3.2 %.

If possible the acceptable error should be defined in National Regulations. If this is not the case then for practical analytical purposes it is suggested that a minimum of 1000 counts are made for each size fraction.

Note 14: *For reliable assessment, the number of points to be counted in each thin-section must exceed the number of particles present, so as to ensure that smaller particles are correctly represented in the assessment. The spacing interval of the virtual square counting grid must be adjusted for each size fraction to cover the entire prepared specimen area (see Table 3). Depending on particle size and thin-section sizes, the total number of particles to be counted may have to be divided over multiple thin-sections* [15].

Table 3 Examples of relative errors after assessment of thin-section(s) for aggregate constituents at five abundance levels, horizontally grouped per size fraction

	Total number assessed		Relative error [%] at abundance level				
	Particles	Counts	50 vol. %	20 vol. %	10 vol. %	5 vol. %	2 vol. %
Alternative 1	*Size fraction 2–4 mm, ~150 particles/~250 counts per section, 2.00 mm grid*						
	600	1000	8	16	24	32	48
	300	500	12	22	32	44	63
	150	250	16	32	44	64	80
	Size fraction 1–2 mm, ~600 particles/~900 counts per section, 0.75 mm grid						
	2400	3600	4	8	12	16	24
	1200	1800	6	12	16	24	36
	600	900	8	16	24	32	48
	Size fraction 0.063–1 mm, ~1800 particles/~2500 counts per section, 0.50 mm grid						
	7200	10000	* <2	*4	*6	*8	*12
	3600	5000	>3	7	10	15	22
	1800	2500	5	10	15	20	30
Alternative 2	*Size fraction 2–4 mm, ~150 particles/~250 counts per section, 2.00 mm grid*						
	600	1000	8	16	24	32	48
	300	500	12	22	32	44	63
	150	250	16	32	44	64	80
	Size fraction 0.063–2 mm, ~1200 particles/~1600 counts per section, 0.50 mm grid						
	4800	6400	<3	6	9	12	20
	2400	3200	4	8	12	16	24
	1200	1600	6	12	16	24	36

*Extrapolated from Fig. B.2 in Annex B

Calculation of modal contents

When all thin-sections have been counted, the modal content, M, in volume percent for each identified lithology, should then be calculated by dividing the sum of counts for each lithology by the total number of counts:

$$M\ [\mathrm{vol}\,\%] = \Sigma \mathrm{lithology}/\Sigma \mathrm{total}\ \times 100\,\%$$

Calculated percentage values should be rounded off to the nearest integer and presented without decimals. All data on lithological groups per size fraction, respective alkali-reactivity potentials according to local experience, modal contents and analytical error should be collated in a table.

6.5 Crushing of Coarse Aggregate Material

Crushing of coarse aggregate is not usually required in preparation for petrographic examination. However, crushing can sometimes be useful for quantitative assessment of bulk coarse aggregate materials >4 mm using thin-section petrography and point counting.

In appropriate cases, absolute and relative errors are strongly reduced by crushing, which is attributable to strict reduction of spread in particle size, as well as the greatly increased number of particles to be counted in the prepared specimen compared to non-crushed sample material.

7 Classification

The procedure described here produces a reliable assessment of concrete aggregate by petrographic identification of the composing mineral and rock constituents.

This can be used to classify the alkali-reactivity potential of the bulk aggregate material, applying criteria based on local (national, regional) experiences, recommendations and specifications, according to the schedule below:

Class I—Very unlikely to be alkali-reactive;
Class II—Alkali-reactivity uncertain;
Class III—Very likely to be alkali-reactive.

In cases of uncertainty, carry out this classification using the following definitions:

Class I aggregate material does not contain known alkali-reactive rock types and/or mineral species in significant amounts;
Class II aggregate material cannot be unequivocally classified as either Class I or Class III;
Class III aggregate material does contain alkali-reactive rock types and/or mineral species in sufficient amount known to cause damage in concrete.

In the cases of Class II and Class III aggregate samples, additionally sub-classify the material according to the siliceous and/or carbonate nature of the potentially reactive constituents, using the following definitions:

Class II-S & **Class III-S** aggregate samples contain particulate constituents judged to be potentially with alkali-silica reactivity (ASR);
Class II-C & **Class III-C** aggregate samples contain particulate constituents judged to be potentially with alkali-reactive carbonates (ACR);
Class II-SC & **Class III-SC** aggregate samples contain both particulate constituents judged to be potentially alkali-silica reactive (ASR) and particulate carbonate constituents judged to be potentially reactive.

Note 15: *The above classification is based on the current recommendations of AAR-0. However, this method would be equally applicable to any future modifications of the AAR-0.*

8 Report

The report should state that the petrographic analysis is compliant with this RILEM method, with any selected options or deviations from the procedure being declared and explained. The report should state explicitly that the observations and acquired results, their interpretation, and drawn conclusions all apply specifically and exclusively to the samples investigated, and are under no circumstances extendable to different samples. Every effort should be made to ensure, though without oversimplification, that persons without background in petrography are able correctly to understand the conclusions of the report.

The following information should be included in the report:

- Name and address of the petrographic laboratory, unique identification number and reporting date, name of the petrographer, signature of the person responsible;
- Name and address of the organization and/or person ordering the assessment;
- Deposit or extraction site name preferably indicated on a map, and if available exact sample location with GPS co-ordinates;
- Sampling collection method, date and person responsible;
- Amount of sample material in kg as received at the laboratory, and delivery date;
- Size fraction of the material, e.g., sand ≤4 mm, gravel >4 mm, and type of material, e.g., natural, processed, crushed, or blended;
- Specification in adequate detail of all sample and specimen preparation techniques applied, and any resulting artifacts observed in the investigated specimen;
- Specification in adequate detail of all assessment procedures applied, equipment, instrumentation, and resulting net analytical error. This includes specifying whether the material has been assessed using macroscopic petrography, or with additional thin-section petrography for identification;
- Number of size fractions, the total number of aggregate particles assessed per fraction and the cumulative sum total for the sample as a whole;
- All relevant observations from petrographic analysis, both macroscopic on unprepared material and on thin-sections using a microscope, including all minerals and rock types that have been identified with their names, and respective alkali-reactivity potentials;
- Where required the modal contents of constituents shall be reported in numbers of counts recorded as well volume or weight percent. Any conversions to/from mass percent or alternative units shall be specified in detail;
- The resulting net sum reactivity classification for the bulk aggregate sample, with explanation in plain text;

- Reference to local recommendations, specifications, earlier reports, or other written sources that were consulted for the present assessment.

Note 16: *Some items listed above (or parts thereof) are subject to availability. If they are available, they should be included as indicated.*

Annex A: Definitions

Systematic rock classification and nomenclature reflects mineral content, structure and texture/fabric as well as occasionally petrogenesis, compliant with nomenclature conventions as published in the literature. The text of this AAR-1.1 procedure adheres to IUGS classification schemes for igneous rocks by Le Maitre et al. [16] and metamorphic rocks by Fettes and Desmons [17], and (for lack of an IUGS-approved document) the British Geological Survey (BGS) classification scheme for sedimentary rocks by Hallsworth and Knox [18].

Andesite

Andesite is a fine-grained volcanic rock of intermediate composition. Potentially alkali-reactive constituents include the high-temperature silica polymorphs cristobalite and tridymite, interstitial glass in the matrix, or devitrified glass, opaline silica or chalcedony from low-temperature alteration.

Argillite, Argillaceous

Argillite is a diagenetically altered or very low-grade metamorphic, very fine-grained rock consisting of >50 vol. % siliciclastic fragments, of which >75 vol. % is smaller than 32 μm, i.e., silt or clay. The rock represents indurated and lithified muds and oozes, typically displays conchoidal fracture but has negligible fissility. With development of cleavage, however, argillaceous rocks grade into slate and higher grade rocks (see entry 'Slate'). An argillite rich in alumina is known as 'pelite', their metamorphic equivalent as 'metapelite'. Potentially alkali-reactive constituents of these rocks are microcrystalline quartz. For the non-metamorphic equivalent of argillite, see entry for 'Mudstone'.

Basalt

Basalt is a fine-grained volcanic rock of mafic composition, chemically slightly poorer in silica than andesite. Potentially alkali-reactive constituents include the high-temperature silica polymorphs cristobalite and tridymite, interstitial glass in the matrix, or devitrified glass, opaline silica or chalcedony from low-temperature alteration.

Chalcedony

Chalcedony is a fine-grained fibrous silica variety, commonly observed in chert/flint, limestone and other sedimentary lithologies. In thin-section, chalcedony has a 'feathery' appearance, with normally length-slow elongation. The alkali-reactivity of chalcedony is generally attributed to its fine grained nature and poor crystallinity, possibly also moganite content.

Chert, Flint

Chert is a non-systematic name for a fine grained sedimentary rock consisting of chalcedonic, opaline, cryptocrystalline and/or microcrystalline silica, occurring as contiguous beds or layers, whereas flint is a nodular equivalent. Mineralogically, chert and flint consist mostly of very fine grained quartz, sometimes with accessory silica minerals chalcedony, opal and moganite. Chert and flint may furthermore contain minor amounts of clay minerals, sulfides, brownish to brick-red stains of iron (oxy-) hydroxides, etc. In fluorescence petrography, chert/flint particles often reveal internal porosity, especially along an outer rim called 'cortex'. The silica in chert/flint generally has a poorly developed crystalline structure, and usually contains hydrous species.

Clays, Clay Minerals

Clays represent a large group of minerals with broadly variable chemical composition, yet sharing a layered structure composed of (alumino-) silicate sheets, similar to micas. Concrete damage from expanding clay minerals is not regarded as alkali-silica reaction, even when alkalis may have contributed to the swelling.

Cristobalite

Cristobalite is a high-temperature SiO_2 polymorph with a lower density than quartz, normally stable >1470 °C until melting at 1730 °C. After cooling down to ambient temperatures, cristobalite may survive in certain felsic volcanic rocks, as recrystallization to α-SiO_2 quartz (stable below 573 °C). Alternatively, cristobalite may form from the devitrification of glass, as e.g., the snowflakes in natural volcanic glass obsidian. Cristobalite is metastable and quick to engage in deleterious ASR.

Cryptocrystalline

Fine-grained rock or mineral texture inhibiting routine petrographic assessment of individual grains in thin-section using an optical microscope, i.e., smaller than ~10 μm. Cryptocrystalline materials do exhibit defined peaks in X-ray diffraction, whereas truly amorphous materials only reveal a wide 'glass bulge' due to increased background noise and absorption.

Diorite

Diorite is a coarse-grained rock type rich in plagioclase, and poor in quartz and K-feldspar. In addition, diorites typically contain pyroxene and other dark minerals. Related rock types are anorthosite and monzonite. Some diorites behave deleteriously due to retrogradation of pyroxene to form amphibole plus micro-crystalline quartz, or to late-stage veining.

Dolomite

Dolomite is a double-carbonate mineral of idealized composition $CaMg[CO_3]_2$. Alternatively, dolomite is the name for a sedimentary rock containing more dolomite than calcite $Ca[CO_3]$ or other non-carbonate minerals. Fine-grained dolomite may be prone to dissolution under high-pH conditions as in concrete (ACR), which however does not cause expansion. Finely disseminated quartz, opal or chalcedony present in dolomite rocks have been shown deleterious by ASR.

Glass (Natural, Industrial)

An amorphous material of natural or industrial origin, with a broad possible range of chemical compositions. Natural glass of volcanic origin occurs as (dark coloured) obsidian of intermediate to mafic composition, light coloured vesicular pumice of felsic composition, or as matrix in andesite, basalt, rhyolite, tuff, and other volcanic rock types. Natural glass of dynamo-metamorphic origin occurs in pseudotachylite. Devitrification occurs by slow recrystallization of the disordered glass structure into cristobalite, e.g., the snowflakes in snowflake obsidian. The alkali-reactivity of natural glass is attributed to its amorphous, non-crystalline structure, rendering it thermodynamically unstable. Especially weathered and partially devitrified/altered/hydrated glasses are known to be prone to deleterious ASR.

Gneiss

Gneiss is a medium to coarse grained high-grade metamorphic rock, with characteristic alternating dark/light coloured bands at millimetre- to metre scale. The banding is caused by differences in mineral content, lighter bands being richer in quartz, feldspars and light micas (e.g., muscovite), darker bands richer in amphibole (e.g., hornblende), dark micas (e.g., biotite) and/or pyroxene (e.g., augite). *Ortho*-gneiss is derived from an igneous precursor, *para*-gneiss from a sedimentary one. The lower-grade equivalent of gneiss is schist, towards higher grade gneiss transforms into granulite. The alkali-reactivity of gneiss is attributed to strained quartz from (tectonic) deformation, and/or poorly (micro-, crypto-) crystalline interstitial quartz.

Grain, Subgrain

An individual granule consisting of a single mineral, e.g., quartz, K-feldspar, plagioclase. Grains of intensely deformed quartz may be divided into multiple subgrains of yet smaller size to reduce local crystal-structural deformation. Thus, the presence of subgraining is considered relevant for alkali-reactivity.

Granite, Granodiorite

Granite is a coarse to medium grained felsic igneous rock of plutonic (=deep-seated) origin containing mainly quartz, K-feldspar and plagioclase, a granodiorite is richer in plagioclase, poorer in K-feldspar. The alkali-reactivity of granite and granodiorite is attributed to strained quartz from deformation (rather common along intrusive margins), or poorly (micro-, crypto-) crystalline quartz.

Granulite

Granulite is a fine grained metamorphic rock with a characteristic granular texture, of higher grade than gneiss, only lower than eclogite. Granulites may show signs of partial melting. The alkali-reactivity of granulite is attributed to strained quartz by deformation, or poorly (micro-, crypto-) crystalline interstitial quartz.

Greywacke

Greywacke is a poorly sorted sandstone, characterised by a very fine grained matrix embedding coarser sedimentary detritus including rock fragments, that may also contain dark organic matter. The alkali-reactivity of greywacke is attributed to very fine grained quartz with a large accessible surface area and/or the presence of poorly (micro-, crypto) crystalline quartz.

Hornfels

Hornfels is a fine-grained contact-metamorphic rock formed at high temperature and shallow depth, characteristically without a foliation or banding (as implied by '-fels'). A generic name for *coarse-grained* isotropic rocks lacking foliation or banding is granofels. The alkali-reactivity of hornfels is attributed to very fine grained quartz with a large accessible surface area and/or poorly (micro-, crypto-) crystalline quartz.

Limestone, Chalk, Marl

Limestone is a consolidated sedimentary rock predominantly composed of calcite $Ca[CO_3]$, possibly with subordinate amounts of dolomite. Depending on origin and geological history, limestones may contain variable amounts of fossil remains. Limestone with 33–67 vol. % clay impurities is known as marl, very pure but little consolidated limestone as chalk (from the Cretaceous formation known as Chalk). Silica from sedimentary detritus, diatom frustules, sponge spiculae or certain types of corals may occur finely dispersed throughout the rock, intercalated with the carbonate as chert or flint, or filling vugs as opal or chalcedony. The alkali-reactivity of limestone is attributed, at least in most cases, to the presence of very fine grained and poorly (micro-, crypto-) crystalline silica.

Microcrystalline

Fine grained rock or mineral texture requiring optical thin-section petrography for reliable identification. In practice, this applies to rocks with grain size less than ~0.1 mm (=100 μm) that cannot be readily identified macroscopically with the unaided eye or a hand lens.

Moganite

Moganite is a little known silica polymorph occurring frequently in chert and flint, and in other lithologies altered at low temperature (e.g., Parisian Basin sandstone, Mogan Formation-Gran Canary basalt). However, its frequent presence in the most common types of concrete aggregate lithologies renders moganite a very plausible contributor to deleterious AAR.

Mudstone

Mudstone is a generic name for a sedimentary rock type from an indurated clay-rich precursor, lacking or with poorly developed fissility. Claystone or shale are generally considered only diagenetically compacted and consolidated, shale showing fissility. Slate does have a defined cleavage and may contain newly formed lowest-grade metamorphic minerals (e.g., chlorite, muscovite) identifiable in thin-section, which are macroscopically recognizable in phyllite (e.g., muscovite, biotite). The alkali-reactivity of some mudstones *sensu lato* is attributed to the presence of fine-grained and/or poorly (micro-, crypto-) crystalline quartz.

Mylonite, Pseudotachylite, Fault Breccia/Gouge, Cataclasite

Mylonite is a dynamo-metamorphic rock with a thinly foliated structure, formed by ductile deformation under geological conditions. The glassy matrix in "pseudotachylite" results from seismic friction-melting and rapid quenching immediately after. Under brittle geological deformation conditions, "fault breccia" forms ("fault gouge" being its finer counterpart). These rock types are alternatively called "cataclasites" The alkali-reactivity of cataclastic rocks is attributed to the presence of strained quartz, glass matrix (pseudotachylite), and/or very fine grained and poorly (micro-, crypto-) crystalline silica from tectonic crushing.

Myrmekite

Myrmekite is one type of symplectitic intergrowth of quartz and feldspar or plagioclase, common in granitic rocks including granites, gneisses, and others. Myrmekite can be recognized as vermicular ('worm-like') intergrowths of quartz embedded in feldspar, but the opposite has been observed too. As the quartz in myrmekite is fine-grained, it could be prone to develop deleterious ASR, if it can be accessed by the concrete pore solution.

Opal, Opaline Silica

Opal is a hydrated form of silica ($SiO_2{\cdot}nH_2O$) with low density and variable water content, which appears glassy and amorphous in thin-section. Opal varieties include opal-AN and opal-CT, which can be distinguished by X-ray diffraction. The silica in opal-AN is amorphous, whereas in opal-CT the structure bears resemblance to silica polymorphs cristobalite and tridymite. Opaline silica is certainly capable of causing ASR damage in field concrete.

Particle

An individual aggregate granule composed of single or multiple mineral grains (excluding subgrains). Particles are composed of multiple grains from different minerals, or from a single mineral representing a mono-mineralic particle. The term 'grain' is only synonymous with 'particle' if a particle consists of one single mineral grain (disregarding inclusions).

Quartz

Quartz is mechanically hard, lacks cleavage and is chemically stable under normal ambient conditions from pH2 to pH9. Its chemical composition is near to pure SiO_2, and its crystal structure is denoted as α-SiO_2. Micro- or cryptocrystalline as well as strained quartz are considered potentially alkali-reactive.

Quartzite

Quartzite is a crystalline rock predominantly comprising quartz grains. Commonly quartzite is a metamorphic rock. Some quartzite was formed by sedimentary processes and can be termed *ortho*-quartzite. The alkali-reactivity of quartzite is attributed to deformed-strained quartz, and/or the presence of very fine grained and poorly (micro-, crypto-) crystalline silica along grain boundaries or interstitial silica.

Rhyolite

Rhyolite is a fine grained to glassy felsic volcanic rock, richer in silica SiO_2 than andesite and basalt (see respective entries). Potentially alkali-reactive constituents include high-temperature silica polymorphs cristobalite and tridymite, interstitial glass in the matrix, or devitrified glass, opaline silica or chalcedony from low-temperature alteration.

Sandstone, Siltstone

Sandstone and siltstone are clastic sedimentary rocks. Constituents reflect mineral composition of the provenance area after sedimentary processing; quartz, feldspar, rock fragments and matrix are used for classification and nomenclature. Grain size of *sand*stone varies from 2 to 0.063 mm, of *silt*stone from 0.063 to 0.002 mm. Sandstones/siltstones are compacted by overburden loading, and are cemented by neogenic minerals, such as quartz, calcite, iron minerals and clay minerals. The alkali-reactivity of some sandstone/siltstone is attributed to clastic constituents that are themselves reactive, to very fine grained quartz and/or the presence of poorly (micro/crypto) crystalline quartz.

Slate

Slate is a very fine-grained, low-grade metamorphic rock with well-defined cleavage formed from a clay-rich sedimentary precursor. Newly formed metamorphic minerals (e.g., chlorite, muscovite) are identifiable in thin-section. The alkali-reactivity of slate is attributed to the presence of very fine-grained (micro/crypto) crystalline quartz.

Tridymite

Tridymite is a high-temperature SiO_2 polymorph, normally stable >870 °C until 1470 °C when it transforms to cristobalite. After cooling down to ambient temperatures, tridymite may survive in certain felsic volcanic rocks, as recrystallization to α–SiO_2 quartz (stable below 573 °C) requires complete reconstruction of the crystal structure and is very slow. Tridymite is metastable and quick to engage in deleterious ASR.

Tuff

Tuff is a 'pyroclastic' rock of volcanic origin, composed of deposited ash particles, consolidated by welding when still hot (welded tuff), or by weathering and alteration at ambient temperature. Ignimbrite is a welded tuff formed from the deposition of particles of pumice, lapilli, glass shards and crystals. The alkali-reactivity of tuff is attributed to the presence of siliceous glass or devitrified glass, cristobalite or tridymite, poorly (micro-, crypto-) crystalline quartz. Opal and/or chalcedony are common products of natural tuff weathering and alteration and also contribute to its alkali-reactivity.

Undulatory Extinction

Certain types of quartz are observed to have 'undulatory extinction' in thin-section petrography, signifying that certain parts of a contiguous individual quartz grain are oriented at an angle to other parts. This quality has been interpreted as an indication that the quartz grain's crystalline structure is strained, and therefore potentially susceptible to deleterious ASR. However, measurement of undulatory extinction angles to predict alkali-reactivity potential has been found unreliable and is now discouraged. For purpose of reference, some geological literature prefers 'undulous' for 'undulatory'.

Annex B: Estimation of Absolute Error in Counting Analysis

Figures B.1 and B.2 are reproduced with permission from Howarth [19].

The graphs enable graphical estimation of absolute analytical error with 95 % confidence limits for total counts (N) in the range 25–5000, for modal percentages ($100 \times n/N$ %; equals vol. %) found by (point) counting assessment of aggregate. In both graphs, the total number of counts is listed along left and right vertical axes

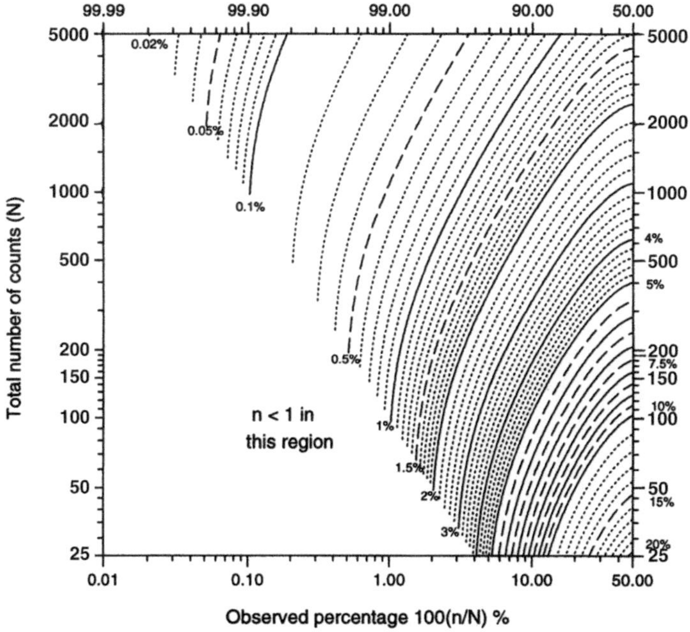

Fig. B.1 *Lower* absolute analytical error, for 95 % confidence limits

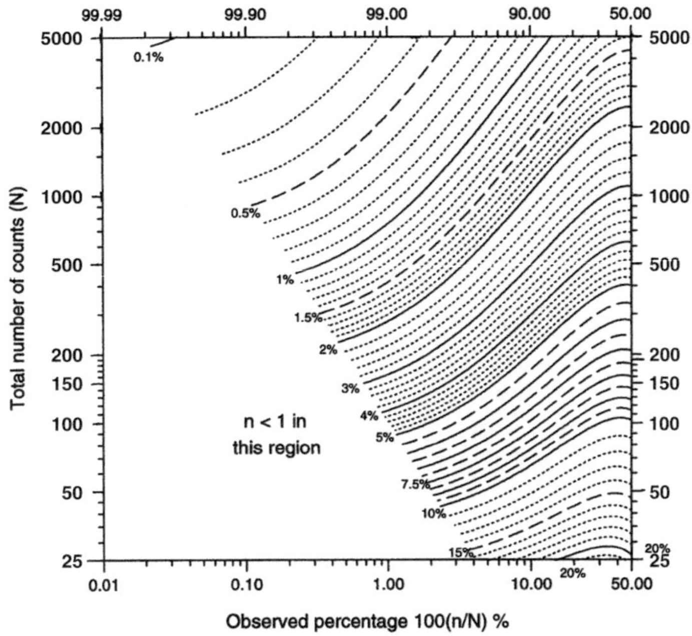

Fig. B.2 *Upper* absolute analytical error, for 95 % confidence limits

(ordinates). Observed percentages (abundances) to 50 vol.% are listed along the lower horizontal (abscissa), observed percentages over 50 vol.% are listed along the upper horizontal. Figure B.1 gives lower bound absolute error, whilst Fig. B.2 gives the upper bound absolute error.

To determine the absolute error for a given observed percentage, first draw a horizontal tie-line connecting the total counts on left and right ordinates. Next, draw a vertical line starting at the observed percentage on the abscissa (lower or upper, as applicable). The curve in the diagram nearest to the point of intersection gives the *absolute error*. Experienced users may prefer using a transparent triangular geometry template of suitable size.

Example: A petrographic analysis with 1000 total counts classifies 20 counts as chert, which represents 2.00 vol. % of the total sample. According to Fig. B.1, the lower absolute error amounts to 0.8 vol. %, whereas the upper absolute error shown in Fig. B.2 amounts to 1.1 vol. %. Thus, the true value lies between (2.0–0.8) = 1.2 vol. % and (2.0 + 1.1) = 3.1 vol. %. By contrast, *relative errors* range from 40 % (=[0.8/2.0] × 100 %) to 55 % (dimensionless!).

Upper/lower error asymmetry decreases for more abundant species. Absolute error can be reduced by increasing the total number of particles assessed.

References

1. Fernandes I et al, (2014): Petrograhic Atlas: Characterisation of aggregates regarding potential reactivity to alkalis, (AAR-1.2) RILEM State-of- the- Art Report, Springer Verlag, Berlin-Heidelberg, DE, to be published.
2. Godart B, De Rooij M, Wood JGM (2013; editors): Guide to diagnosis and appraisal of AAR damage to concrete in structures. Part 1: Diagnosis (AAR 6.1). RILEM State-of-the-art Reports. Springer Verlag, Berlin-Heidelberg/DE: pp87.
3. EN 932-1 (1997): Tests for general properties of aggregates – Part 1: methods for sampling. CEN, Brussels.
4. ISO 565 (1990) (R06): Test sieves – Metal wire cloth, perforated metal plate and electroformed sheet – nominal sizes of openings. International Standards Organization (ISO), Geneva/CH. Standards catalogue TC24/SC8.
5. ISO 3310-1 (2000): Part 1: Test sieves – Technical requirements and testing. International Standards Organization (ISO), Geneva/CH. Standards catalogue TC24/SC8, Corrigendum (2004).
6. ISO 3310-2 (2013): Part 2: test sieves of perforated metal plate. International Standards Organization (ISO), Geneva/CH. Standards catalogue TC24/SC8.
7. ASTM E11 (2009): Standard specification for woven wire test sieve cloth and test sieves. American Society for Testing & Materials, West Conshoshocken, Pennsylvania, USA.
8. EN 933-2 (1996): Tests for geometrical properties of aggregates – Part 2: Determination of particle size distribution - Test sieves, nominal size of apertures. CEN, Brussels.
9. CUR-Recommendation 89 (2008): Measures to prevent concrete damage by the alkali-silica reaction. Official English translation, 2nd revised edition. Centre for Civil Engineering Research and Codes, Gouda, The Netherlands: pp 48.
10. ASTM D75 (2003): Standard practice for sampling aggregates. American Society for Testing & Materials, West Conshoshocken, Pennsylvania, USA.
11. Gy PM (1998): Sampling for analytical purposes. John Wiley & Sons, New York: pp 153.

12. EN 932-2 (1999): Tests for general properties of aggregates. Methods for reducing laboratory samples. CEN, Brussels.
13. BS 812-104 (1994): Testing aggregates- Method for qualitative and quantitative petrographic examination of aggregates. British Standards Institution, London, UK.
14. ASTM C295 (2012): Standard guide for petrographic examination of aggregates for concrete. American Society for Testing & Materials, West Conshoshocken, Pennsylvania, USA.
15. Howarth RJ, and French WJ (1998): A statistical study of aggregate testing data with respect to engineering judgment. In: Latham, JP (editor): Advances in aggregates and armourstone evaluation. Geological Society London, Engineering Geology Special Publications: 169–183.
16. Le Maitre RW, Streckeisen A, Zanettin, B, Le Bas MJ, Bonin P, Bateman P, Bellieni G, Dudek A, Efremova S, Keller J, Lameyre J, Sabine PA, Schmidt R, Sørensen H, Woolley AR (2005): Igneous rocks: a classification of and glossary of terms. Recommendations of the International Union of Geological Sciences (IUGS), Subcommission on the Systematics of Igneous Rocks. 2nd edition. Blackwell Scientific, Oxford: pp256.
17. Fettes D, Desmons J (2007): Metamorphic rocks: a classification and glossary of terms. Recommendations of the International Union of Geological Sciences (IUGS), Subcommission on the Systematics of Metamorphic Rocks. Cambridge University Press, Cambridge: pp258.
18. Hallsworth CR, Knox RWO'B (1999): BGS rock classification scheme (3): Classification of sediments and sedimentary rocks. British Geological Survey Report (RR 99-03): pp44.
19. Howarth RJ (1998): Improved estimators of uncertainty in proportions, point counting, and pass-fail test results. American Journal of Science (298): 594–607.

RILEM Recommended Test Method: AAR-2—Detection of Potential Alkali-Reactivity—Accelerated Mortar-Bar Test Method for Aggregates

1 Foreword

This draft method was originally prepared by RILEM TC 106-AAR (Alkali-Aggregate Reaction—Accelerated Tests) as TC 106-2, and has been revised by RILEM TC 191-ARP (Alkali-Reactivity and Prevention—Assessment, Specification and Diagnosis) and RILEM TC 219-ACS (Alkali-Aggregate Reactions in Concrete Structures) as AAR-2. It is based on the South African NBRI (National Building Research Institute) accelerated test method [1].

2 Scope

This test method is intended to determine rapidly the potential alkali-reactivity of aggregates through the evaluation of the expansion of mortar-bars immersed in NaOH solution at elevated temperature, as specified in the method. Alternative methods are described, depending on the mortar-bar size used: AAR-2.1 (285 × 25 × 25 mm) or AAR-2.2 (160 × 40 × 40 mm). The test may also be used in experiments to assess the pessimum behaviour of reactive aggregates (See Annex—A2). The test is not suited for the determination of the potential alkali-reactivity of aggregates containing more than 2 % (by mass) of porous chert and flint (See Annex—A3).

3 Significance and Use

This test method provides a means of screening aggregates for their potential alkali-reactivity. It may be especially useful for aggregates that react slowly or produce expansion late in the reaction such as granite, rhyolite, andesite, volcanic

P.J. Nixon and I. Sims (eds.), *RILEM Recommendations for the Prevention of Damage by Alkali-Aggregate Reactions in New Concrete Structures*, RILEM State-of-the-Art Reports 17, DOI 10.1007/978-94-017-7252-5_4

tuff, gneiss, quartzite, hornfels, phyllite, sandstone, greywacke, shale, tillite, non-porous chert and flint, diatomite, quartz-bearing dolomitic limestones and quartz-bearing calcitic dolostones. It is recommended that a petrographic examination of the aggregates should accompany the accelerated test method (See Annex—A3).

4 Principle

In this method prisms are moulded from mortar prepared with the aggregate to be tested. Two alternative prism sizes are recommended. The prisms are demoulded after 24 ± 2 h and their initial length measured. The specimens are then placed in water, transferred to an oven at 80 ± 2 °C for 24 h, removed from the water and the length measured immediately before the temperature has dropped substantially (zero reading). The specimens are immediately placed in containers with a 1 M NaOH solution already at 80 ± 2 °C, the containers sealed and placed in an oven at 80 ± 2 °C (subsequent 14 days). Length measurements are taken periodically.

5 Apparatus

5.1 Sieves

A set of sieves conforming to series A of ISO 6274 [2], having square apertures of 4 mm, 2 mm, 1 mm, 500 μm, 250 μm and 125 μm. Alternatively, the equivalent sieves of series B (4.75 mm, 2.36 mm, 1.18 mm, 600 μm, 300 μm and 150 μm) or C (5 mm, 2.5 mm, 1.25 mm, 630 μm, 315 μm and 160 μm) can be used.

5.2 Balance

A balance capable of weighing 1000 g with an accuracy of 1 g.

5.3 Measuring Cylinders

Graduated in ml, with a capacity of 500 ml.

5.4 Mixer, Paddle and Mixing Bowl

Mixer, paddle and mixing bowl, as used in the procedure for testing the strength of cement with plastic mortar.

5.5 Flow Table

Flow table consisting essentially of a circular rigid table top, with a diameter of about 300 mm, that can be raised vertically by means of a cam, and dropped through a nominal height of 10.0 mm (e.g. conforming to EN 1015-3 [3]).

5.6 Mould for Flow Test

A frustum of a cone with height 60 mm, base diameter 100 mm and top diameter 70 mm (e.g. conforming to EN 1015-3 [3]).

5.7 Tamper for the Flow Table

Tamper, made of non-absorbent, non-abrasive, non-brittle material and with a circular cross-section of 40 mm in diameter. A convenient length is 230–300 mm. The tamping face shall be flat and at right angles to the length of the tamper. The tamper mass shall be 0.250 kg (e.g. conforming to EN 1015-3 [3]).

5.8 Prism Moulds

Moulds, providing for prisms with a nominal length of 285 mm and a cross section of 25 × 25 mm (option AAR-2.1) or nominal length of 160 mm and a cross section of 40 × 40 mm (option AAR-2.2). The end plates of the moulds must have threaded holes in the centres to take stainless steel pins of 6 mm diameter and 20 mm length used for length measurements.

5.9 Tamper for the Prism Moulds

Tamper, made of non-absorbent, non-abrasive, non-brittle material and with a cross-section of 13 × 25 mm. A convenient length is 120–150 mm. The tamping face shall be flat and at right angles to the length of the tamper.

5.10 Length Comparator

Consisting of:

- an apparatus to measure the length of the specimens conveniently and rapidly;
- a high-grade dial micrometer, or other measuring device, graduated to read in 1.0 μm units, accurate to within 2.0 μm in any 20 μm range, and within 5 μm in any 0.25 mm range. The measuring range shall allow for small variations (±10 mm) from the nominal gauge length of the specimens;
- an Invar® reference bar (or similar) of the same nominal length as the specimens for checking the measuring device, before and after each set of readings.

5.11 Containers

Rigid containers for the test specimens made of plastics or other material resistant to corrosion by a solution of sodium hydroxide at a temperature of 80 °C for a prolonged period of time. Each container must be of such dimension and shape to accommodate at least three specimens and must be provided with lids or other suitable means to prevent loss of moisture by leaking or evaporation. The prisms must be positioned and supported in such a way that the solution has access to the whole of the bar. It should further be ensured that the specimens do not touch each other or the sides of the container. The specimens, if stood upright in the solution, shall not be supported by the steel pins. A suitable arrangement is shown in Figs. 1 and 2.

5.12 Storage

A cabinet or moist storage room maintained at a temperature of 20 ± 1 °C and a relative humidity not less than 90 % (e.g. conforming to EN 196-1 [4]).

Fig. 1 One suitable container, showing the internal framework for holding the mortar-bars (note the perforations in the base, to accommodate the measuring studs)

Fig. 2 As Fig. 1, but showing the set of mortar-bars held in the internal framework

5.13 Oven

An oven or room of suitable size to accommodate the required number of containers maintained at a temperature of 80 ± 2 °C.

5.14 Crusher

A laboratory-type crusher of suitable size and design capable of crushing aggregate to the prescribed size fractions.

6 Reagents and Materials

6.1 Water

Distilled or deionised water.

6.2 Sodium Hydroxide Solution (NaOH)

Each litre of sodium hydroxide solution shall contain 40.0 g of NaOH dissolved in 900 ml of water and, after cooling to about 20 °C, it shall be diluted with additional distilled or deionised water to obtain 1.0 litre of solution. The concentration of the solution shall lie between 0.99 and 1.01 M. A new solution shall be prepared for each series of tests.

6.3 Cement

An ordinary Portland cement CEM I [5] or ASTM type I [6] with a minimum Na_2O equivalent (Na_2O + 0.658 K_2O) of 1.0 % shall be used. The specific surface of the cement, when measured according to the air permeability method (e.g. EN 196-6 [7]), shall be greater than 450 m^2/kg. The autoclave expansion, determined according to ASTM test method C151 [8] shall be less than 0.20 %. Alternatively the MgO soundness can be evaluated through a Le Chatelier test (e.g. EN 196-3 [9]) and the increase in separation of indicators ends shall be 0 mm. Check the cement to see if any lumps are present. If these cannot be crushed between your fingers, sieve the cement on a 250 μm sieve.

Note 1: *A suitable reference cement is available from Norcem A.S (Norway). References for this cement (CEM I 42.5 R) are reported in RILEM Recommendation AAR-0, including properties.*

7 Procedure

7.1 Preparation of the Aggregate Sample

The quantity of sample delivered to the laboratory should be in accordance with the sampling procedures specified in RILEM Recommended Test Method AAR 1.1 (Detection of Potential Alkali-Reactivity of Aggregates: Petrographic Method). The material to be tested shall consist of particles ≤4 mm and ≥125 μm, graded according to Sects. 7.1.1, 7.1.2 and 7.1.3, washed and dried as indicated in Sect. 7.1.5 (see Annex A7.1).

7.1.1 Natural Fine Aggregate

The natural material proposed for use as fine aggregate in concrete shall be tested in the grading as submitted, except that particles retained on a 4 mm screen or equivalent shall be included after crushing. Also particles passing 125 μm shall be completely removed.

Note 2: *The particles retained on 4 mm screen or equivalent shall be removed only if petrography has confirmed that they are not different from the particles passing through.*

7.1.2 Crushed Fine Aggregate

The crushed material proposed for use as fine aggregate in concrete shall be tested in the grading prescribed in Table 1.

Table 1 Grading requirements

Sieve Size		Mass (%)
Passing	Retained	
4 mm	2 mm	10
2 mm	1 mm	25
1 mm	500 μm	25
500 μm	250 μm	25
250 μm	125 μm	15

See Annex A7.1.2

7.1.3 Coarse Aggregate

The material proposed for use as coarse aggregate in concrete shall be processed by crushing and sieving to produce a graded sample in accordance with the requirements prescribed in Table 1. This sample shall be representative of the composition of the coarse aggregate as proposed for use.

Note 3: *Coarse aggregate crushed to sand size may give increased expansion, owing to the increased surface exposed upon crushing. Therefore, if coarse aggregate tested by this method is found to be potentially reactive, tests should be performed on concrete specimens, according to RILEM Recommendation AAR-0 to verify the results found with the mortar samples.*

7.1.4 Fine and Coarse Aggregate

The material proposed for use as fine and coarse aggregate in concrete shall be separated by sieving on a 4 mm sieve or equivalent. Fine and coarse aggregate shall be tested separately as indicated in clauses Sects. 7.1.1, 7.1.2 and 7.1.3.

7.1.5 Final Aggregate Preparation

After the fine or coarse aggregate has been processed and washed to remove adhering dust or fine particles (<125 μm) from the aggregate, dry the fractions at 100–110 °C for 16 h and cool to ambient temperature. Unless used immediately, store each fraction individually in a clean container provided with a tight-fitting cover.

7.1.6 Combining the Aggregates

If the test method is to be used in an experimental programme to assess the pessimum behaviour of a reactive aggregate, undertake a petrographic examination of the test sample according to RILEM Recommended Test Method AAR 1.1 (Detection of Potential Alkali-Reactivity of Aggregates: Petrographic Method). This examination should be sufficiently detailed to identify all the rock types or mineral constituents that compose the test sample. The results from the petrographic examination could then be used to establish a suitable test protocol (See Annex A7.1.6) such that any potentially damaging pessimum behaviour is identified.

7.2 Conditioning

The temperature of the moulding room, apparatus, dry materials, mixing water shall be maintained at 20 ± 2 °C. The temperature of the cabinet or moist storage room shall be maintained at 20 ± 1 °C. The relative humidity of the moulding room and of the cabinet or moist storage room shall not be less than 50 and 90 % respectively.

7.3 Proportioning of Mortar

Proportion the dry materials for the test mortar using 1 part of cement to 2.25 parts of aggregate by mass. The minimum quantity of dry materials to be mixed at one time for making three 25 × 25 × 285 mm specimens (option AAR-2.1) is 400 g of cement and 900 g of dry aggregate. To make three 40 × 40 × 160 mm specimens (option AAR-2.2) a minimum amount of 600 g of cement and 1350 g of dry aggregate will be necessary.

Use a free water/cement ratio of 0.47 by mass, where the free water is the water available for hydration of cement and for the workability of the fresh mortar. The total water added to the mix is the free water plus the water absorbed by the aggregate to bring it to a saturated surface dry condition. The water absorption can be measured by the EN 1097-6 method [10].

Measure the workability of the mortar on a flow table (e.g. EN 1015-3 [3]). The flow value (mean diameter of the mortar) shall be in the range of 205–220 mm and, in the cases of mixes with a flow initially less than 205 mm, use a superplasticiser (not one combined with an air entraining agent) in order to achieve the required flow.

Note 4: *The water content of the superplasticizer should be included in the free water content.*

Note 5: *Alternatively, the ASTM C1437* [11] *method may be used, when the flow range should be 105–120 %.*

7.4 Moulding Test Specimens

Make at least three specimens for every aggregate to be tested.

Prepare the specimen mould with a suitable releasing agent that will not affect the setting of the cement or leave any residue that will inhibit the penetration of water into the specimen. Mix the mortar in accordance with the procedure for testing the strength of cement with plastic mortar (e.g. EN 196-1 [4]). If a superplasticiser is used, add it at the end of this mixing procedure and then mix for a further 60 s. Start moulding the specimens not more than 3 min after completion of the mixing of the mortar batch. Fill the moulds with two approximately equal layers, each layer being compacted with the suitable tamper. Work the mortar along the surface of the mould with the tamper until a homogeneous specimen is obtained, making sure that the mortar is fully pushed under the reference inserts before a second layer is placed into mould. After the top layer has been compacted, cut off the mortar flush with the top of the mould and smooth the surface with a few strokes of the trowel.

Note 6: *Particular care should be taken to attain a consistent compaction of the mortar, as the degree of compaction greatly influences the degree of expansion, the better the compaction the lower the expansion.*

7.5 Initial Curing and Measurement

Place the moulds in the moist cabinet or in the moist storage room (20 ± 1 °C and not less than 90 % RH) for a period of 24 ± 2 h. Remove the specimens from the mould and, while they are being protected from loss of moisture, properly identify each specimen in such a way that they, when subsequently measured, are placed in the measuring equipment in the same manner. Make and record the length (Li) and all subsequent measurements to the nearest 0.001 mm.

Place the specimens made with each aggregate sample in a storage container with sufficient distilled or deionised water, at room temperature, to immerse them totally. Seal and place the containers in an oven at 80 ± 2.0 °C for a period of 24 h.

Remove the containers from the oven one at a time. Remove other containers only after the bars in the first container have been measured and returned to the oven. Remove the bars one at a time from the water and dry their surfaces with a towel or cloth paying particular attention to the two metal pins. Take the zero measurement of each bar (Lo) immediately after drying and read as soon as possible after the bar is in position. Complete the process of drying and measuring within 15 s of removing the specimens from the water.

The measuring device should be checked with the reference bar prior to and after measurement of each set of specimens.

7.6 Final Storage and Measurement

Place the specimens made with each aggregate sample in a container with sufficient 1 M NaOH, preheated at 80 ± 2.0 °C, totally to immerse the specimens. The recommended volume proportion of sodium hydroxide solution to mortar-bars in a storage container shall be 4 ± 0.5 times the volume of the mortar-bars. Seal the container, mark the level of the liquid on the outside and return it to the oven.

Take subsequent measurements (Ln) of the specimens periodically, with a reading after 24 h of immersion in the NaOH solution and at least three intermediate readings before the final reading at 14 days (e.g. 3, 6 and 9). If so desired, measurements may be taken at 24-h intervals and may be continued beyond 14 days from the zero measurement (e.g. 21, 28 and 56 days). All measurements should be taken at approximately the same time each day. The measuring procedure is identical to that described in clause 7.5 and the specimens are returned to their container after each measurement. If moisture is lost from the container by evaporation replenish with distilled or deionised water.

In some cases, the solution may become cloudy due to the presence of alkali-silica gels.

7.7 *Safety Measures*

Reliable safety precautions should be taken and suitable personal protective equipment should always be used to avoid the hazards of the hot alkaline solution such as severe burns and injury to unprotected skin and eyes.

8 Calculation and Reporting of Results

8.1 *Expansion*

The linear expansion of each specimen is obtained by calculating the difference between the length of the specimen at each period of measurement (Ln) and the zero measurement (Lo), to the nearest 0.001 % of the effective length, as follows

$$Expansion, \% = 100 \times (Ln - Lo)/Gauge\,length$$

where:

Lo = measurement of specimen just before immersion in sodium hydroxide solution (zero measurement),
Ln = reading taken at each period of storage in sodium hydroxide solution, n being the number of days counted from the zero measurement,
Gauge length = distance between inner ends of the metal pins, measured to the nearest of 0.1 mm.

Lo is taken after 1 day of storage of mortar specimens in water bath at 80 ± 2.0 °C and just before their immersion in sodium hydroxide solution.

Ln is the measurement referred to the days of immersion in sodium hydroxide solution at 80 °C, starting immediately after the zero measurement. Then the measurement Ln = L_{14}, taken after 14 days of immersion, corresponds to the measurement taken 16 days after casting.

The gauge length, corresponding to the distance between the inner ends of the metals pins, has to be accurately evaluated after demoulding (when making the initial measurement Li), taking into account the length of the metal pins, these last being measured to the nearest of 0.1 mm.

8.2 Expansion Recording and Further Examination

Record and report to the nearest 0.01 % the average expansion of the specimens for a given period. If a specimen breaks during the test, the test will be considered valid provided the two remaining specimens do not differ from each more than the values given below. For average values of expansion greater than 0.10 %, the repeatability is considered satisfactory if the expansion of each specimen is within 10 % of the average value, with a confidence level of 95 % (i.e. the probability of one case out of 20 of an incorrect result). For average values of expansion less than 0.10 %, the repeatability is considered satisfactory if, for each specimen, the standard deviation from the average value is within 0.01 %. If these values are exceeded, deem the test invalid and repeat the test.

After the final measurement, the specimens shall be examined and any relevant features recorded. Warping, if observed, shall be measured on 3 moulded surfaces by placing the specimens on a plane surface, with curved ends facing downwards, and measuring the maximum separation between the specimen and the surface to the nearest 0.2 mm. Location, type and pattern of any cracking should also be recorded.

Note 7: *If the mortar-bars exhibit an expansion that is judged to be deleterious, a petrographic examination of the interior of the bars that have expanded most should be carried out together with an examination of the internal crack pattern to confirm that the cause of expansion is likely to be an alkali-silica reaction* [12]. *If it has been concluded from the expansion results and supplementary examination of the prisms that a given aggregate should be considered potentially alkali-reactive, additional studies using the RILEM Recommended test methods may be appropriate to develop further information on its potential alkali-reactivity and to evaluate the effect of coarse aggregate, different aggregate gradings and different alkali contents of the concrete.*

Note 8: *Recently, an interpretation of the accelerated mortar-bar test results with a kinetic-based method has been proposed* [13–15]. *It provides a criterion for defining potentially alkali-reactive aggregates that is an alternative to the percent expansion criterion (see Annex* A8*).*

9 Test Report

The following information shall be given in the report:

- identification and source of the aggregate sample and reference to petrographic analysis, if available; date of and state at delivery at the laboratory;
- type and maximum size of the aggregate;
- type of processing undertaken on the aggregate sample in the laboratory (washing, drying, crushing, sieving etc.);
- grading of the aggregate as used in the test;
- identification and source of the Portland cement;
- alkali content of the cement expressed as equivalent sodium oxide (% Na_2O + 0.658 % K_2O);
- autoclave expansion or Le Chatelier value of the cement;
- Blaine fineness of the cement;
- workability of mortar (flow);
- type and content of superplasticiser (if used to achieve the mortar workability in the required range), and its contribution to the alkali content of the mortar mix;
- size of the mortar-bars. i.e. option AAR-2.1 or option AAR-2.2;
- tested combinations of the aggregate with a reference non-reactive fine material (if the pessimum behaviour has been investigated);
- initial expansion of the bars after 24 h of storage in water at 80 °C;
- each single value and average percentage length change after each measurement of the specimens;
- a graph of the percentage length change vs time from the zero reading to the end of the 14-day period of immersion in NaOH solution;
- results of any warping measurements of the specimens;
- any significant features revealed by examination of the specimens and the sodium hydroxide solution during and after the test.

Annex

(Comments relate to clauses as numbered in method)

A2 Scope

With some reactive aggregates it has been found that there is a proportion of reactive constituents in the aggregate that leads to a maximum expansion. This proportion is called the "pessimum" content and the relationship between expansion and reactive constituents content is called the "pessimum behaviour" of the reactive aggregate.

A3 Significance and Use

It is recommended to start all screening of aggregates for their potential alkali-reactivity with a petrographic examination of the aggregate. On the basis of the results of the petrographic examination, a decision can be made with respect to further testing. If aggregates contain more than 2 % (by mass) of porous chert and flint, further testing by means of the accelerated mortar-bar test is not recommended. It is reported (see [16, 17]) that application of the accelerated mortar-bar test to reactive aggregates containing porous chert and flint sometimes gives rise to misleading results and inappropriate approval of such aggregates.

A7.1 Preparation of the Aggregate Sample

125 μm is the lowest particle size considered in the aggregate sample, both for crushed or natural fines. This is in the light of avoiding discrepancies between the reactivity diagnoses of certain aggregates obtained under different test temperatures. Indeed, at higher temperatures (80 °C), finer particles (e.g. 63 μm or less) may develop a pozzolanic activity that may balance the alkali-reactivity of aggregates, in contrast with what happens at 38 °C (see [18]).

A7.1.2 Crushed Fine Aggregate

Alternatively the equivalent sieves of series B and C of ISO 6274 [2] could be used.

A7.1.6 Combining the Aggregates

If the aggregate sample is composed of:

(a) particles of a single rock, sand, gravel or mineral type that has previously been shown not to have an associated pessimum behaviour and it is intended to use this aggregate unblended; or,

(b) the reactive constituent in a single rock, sand, gravel or mineral type is found at concentrations outside the pessimum limits previously established for the reactive constituent type and it is intended to use this aggregate unblended,

then only a mortar with an aggregate component composed wholly of the test aggregate needs to be tested.

Alternatively, if an aggregate sample is:

(a) composed wholly or in part of a rock, sand, gravel or mineral type which has particles or mineral constituents that have an associated pessimum proportion and the reactive constituents are found in concentrations that are deemed to be

Table A.1 Suggested aggregate proportions

Combination	Aggregate to be tested (%)	Reference non-reactive fine material (%)
I	100	0
II	50	50
III	25	75
IV	15	85
V	5	95
VI	0	100

potentially reactive, or it is either unknown or not prescribed where the "safe" limits lie; or,

(b) if it has not been established whether the rock or mineral constituent type has or does not have an associated pessimum proportion; or,

(c) if the sample is a blend of aggregate types where the behaviour of the aggregates when combined is unknown; or,

(d) if a petrographic analysis is not available,

then several combinations of test aggregate and non-reactive fine material will need to be tested, as indicated in the following paragraph, in order to identify any pessimum behaviour.

The aggregate to be tested shall be mixed with a reference non-reactive fine material (natural sand or crushed rock) with the same grading as the sample under test. The material used for this purpose shall give an expansion of less than 0.05 % by this test method (See Note A9) and shall not exhibit a pessimum behaviour. It should be preferably a non-siliceous material and previously evaluated. Some suggested proportions of test aggregate and reference non-reactive fine material are shown in Table A.1. As preliminary tests, only three combinations should be used (e.g. I, II and IV in Table A.1). Special aggregates may require some other levels to make sure that any possible pessimum behaviour is identified.

Note A9: *This expansion limit is deliberately more restrictive than is generally applied to identify non-reactive aggregates, to ensure that the non-reactive reference aggregate has minimal effect on the results of this test.*

A8 Calculations and Reporting of Results

Criteria for interpreting the results of this test are given in AAR-0, including different maximum percentage expansion criteria for the AAR-2.1 and AAR-2.2 alternative methods.

Recently, an interpretation of the accelerated mortar-bar test results with a kinetic-based method has been proposed. This provides a criterion for defining potentially alkali-reactive aggregates that is an alternative to the maximum percentage expansion criterion.

Correlation of expansion data with a kinetic equation provides a kinetic parameter, ln k, assumed as a reactivity criterion, that is capable of differentiating between reactive and innocuous aggregates regardless of the expansion values obtained at 14 days of immersion of mortar bars in 1 M NaOH solution at 80 °C. This kinetic equation is

$$E_t(\%) = 1 + E_{t0}(\%) - \exp\left[-k(t - t_0)^M\right] \quad (A.1)$$

where E_t is the percent expansion at the time t (days) of immersion of mortar bars in NaOH solution, E_{t0} is the percent expansion at the time t_0 (days) of immersion in NaOH solution, t_0 is the induction period (days) before growth-based kinetics begin, k is a rate constant which combines the effect of nucleation, multidimensional growth, geometry of reaction products and diffusion, and M is an exponential term related to the form and growth of the reaction products. Further details on the development of Eq. (A.1) are given in Johnson and Fournier [13] and Johnson et al. [14].

Solving the logarithmic form of this equation by a least-squares fit yields the reactivity parameter, ln k, as the intercept of the regression line.

A tentative value of three days for t_0 has been proposed in the above references [13, 14], to correlate the accelerated mortar-bar results with Eq. (A.1). Alternatively t_0 may be determined as the value that will give that maximum expansion rate constant, k [15].

References

1. Oberholster, R E, Davies, G, An accelerated method for testing the potential alkali-reactivity of siliceous aggregates, Cement and Concrete Research, 16 (2), 1986, 181–189.
2. ISO 6274:1982, Concrete—Sieve analysis of aggregates.
3. EN 1015-3:1999 (eg BS EN 1015-3), Methods of test for mortar for masonry—Determination of consistence of fresh mortar (by flow table).
4. EN 196-1:2005 (eg BS EN 196-1), Methods of testing cement—Determination of strength.
5. EN 197-1:2011 (eg BS EN 197-1), Cement—Composition, specifications and conformity criteria for common cements.
6. ASTM C150-05, Standard specification for Portland cement.
7. EN 196-6:2010 (eg BS EN 196-6), Methods of testing cement—Determination of fineness.
8. ASTM C151- 05, Standard test method for autoclave expansion of Portland cement.
9. EN 196-3:2005 (eg BS EN 196-3), Methods of testing cement—Determination of setting time and soundness.
10. EN 1097-6:2000 (eg BS EN 1097-6), Tests for mechanical and physical properties of aggregates—Determination of particle density and water absorption.
11. ASTM C1437- 07, Standard test method for flow of hydraulic cement mortar.
12. Report of a Working Party, The Diagnosis of Alkali Silica Reaction, 2nd Edition, 1992, BCA (British Cement Association), Slough (now Crowthorne), Uk.
13. Johnston, D, Fournier, B, A kinetic-based method for interpreting accelerated mortar bar test (ASTM C1260) data, Proceedings, 11th International Conference on Alkali-Aggregate Reaction, Québec, 2000, 355–364. Centre de Recherche Interuniversitaire sur le Béton (CRIB), Laval, Canada.

14. Johnston, D, Stokes, D, Surdahl, R. A kinetic-based method for interpreting ASTM C1260, Cement, Concrete and Aggregates, CCAGDP, 22 (2), 2000, 142–149.
15. Stokes, D. Concerning the use of expansion data from ASR testing, Proceedings, Marc-André Bérubé Symposium on Alkali-Aggregate Reactivity in Concrete, part of 8th CANMET/ ACI International Conference on Recent Advances in Concrete Technology, Montréal, Canada, 2006, 93–101.
16. Soers, E. The alkali-aggregate reaction in Belgium, Proceedings of the Advanced Seminar on Alkali-Aggregate Reaction—the European Dimension, Queen Mary and Westfield College, University of London, London, 1990, 147–160.
17. STAR Project, Standard tests for alkali-reactive rocks, Final Report, European Commission, DG XII, contract n° SMT4-CT96-2128, 1998, 148 pp.
18. Pedersen, B, Mortsell, E, Jensen, V. Effects of alkali-reactive crushed fillers on expansions due to ASR, Proceedings of 12th ICAAR in Concrete, Beijing, China, October 2004, 764–772.

RILEM Recommended Test Method: AAR-3—Detection of Potential Alkali-Reactivity—38 °C Test Method for Aggregate Combinations Using Concrete Prisms

1 Foreword

The former version of this method was initially developed by RILEM following an international trial. This trial showed that the method can reliably differentiate reactive and non-reactive combinations for a range of aggregate compositions from around the world. It was published as a draft for comment in Materials and Structures in 2000 [1]. It has been further developed and amended by RILEM following comments on the draft and the experience of the European PARTNER programme [2] and the storage conditions are now similar to those given in ASTM C1293 [3]. It is one of a suite of test methods for aggregate reactivity developed by RILEM, the combined use of which, for aggregate assessment, is explained in AAR-0.

2 Scope

The method covers the measurement of expansion produced by alkali-aggregate reaction of concrete prisms stored in an environment which accelerates the reaction. It enables the testing of the potential alkali-reactivity of aggregate combinations (Application 1: AAR-3.1). Additionally, the method allows the determination of the alkali threshold of an aggregate combination (Application 2: AAR-3.2).

The method has been developed for *normal weight* aggregates and is not necessarily applicable for lightweight *(oven dry particle density less than 2000 kg/m³)* or heavy weight aggregates *(oven dry particle density greater than 3500 kg/m³).* Additional information concerning the testing methodology is given in Annex A.

P.J. Nixon and I. Sims (eds.), *RILEM Recommendations for the Prevention of Damage by Alkali-Aggregate Reactions in New Concrete Structures*,
RILEM State-of-the-Art Reports 17, DOI 10.1007/978-94-017-7252-5_5

3 Principle

In AAR-3.1, concrete test prisms are prepared from the aggregate combination under test (either the standard test with fixed coarse:fine aggregate proportions, or the alternative test with variable aggregate proportions) and are stored in warm (38 °C), humid conditions for 12 months to promote any alkali-aggregate reaction. Measurements are made at periodic intervals to determine whether any expansion has occurred. To promote further any potential for reaction, the prisms are made with a relatively high content of Portland cement with a high alkali content such that the alkali level in the concrete is 5.5 kg sodium oxide equivalent per cubic metre of concrete. Sodium hydroxide is added to the mix when necessary to enhance the alkali level.

In order to identify the effect of specific aggregate combinations and any pessimum effects, there are options to test the coarse and fine test aggregates together or either in combination with a non-reactive material.

For AAR-3.2, the alkali content is adjusted between 2 and 5 kg/m^3, using sodium hydroxide added to the mix when necessary to enhance the alkali level.

4 Apparatus

4.1 Sieves

Aperture sizes (mm) 22.4, 16.0, 8.0, 4.0, 2.0, 1.0, 0.5, 0.25 and 0.125 or ASTM equivalents [4].

4.2 Moulds

Three steel moulds suitable for casting concrete prisms of lengths 250 ± 50 mm and cross-section 75 ± 5 mm. The moulds shall have the facility for casting stainless steel reference studs into the mid-points of the end faces of the prisms.

4.3 Length Comparator

The length comparator should be such as to accommodate the shape of the reference studs in the prism and shall incorporate a gauge mounted rigidly in a vertical orientation. The graduation of the gauge shall not be greater than 0.002 mm and the error throughout the range of traverse shall be no more than 0.005 mm. Horizontally mounted comparators are not suitable for this test method.

4.4 Standard Length Gauge

This shall consist of an Invar® rod with ends machined to accommodate the reference studs.

4.5 Weighing Device

Suitable for weighing specimens to the nearest 0.1 g.

4.6 Concrete Mixing and Casting Equipment

This shall be suitable for mixing batches (EN 480-1 [5]) and vibrating into the moulds (EN 12390-2 [6]) or ASTM C 192/C192 M [7].

4.7 Specimen Storage

The specimens shall be stored in roughly cylindrical containers (e.g. plastic pails) with airtight lids. The containers shall be of sufficient size to store three specimens vertically without being in contact with each other or the sides of the container. A perforated rack shall be placed in the bottom of the container so that the prisms are 40–50 mm above the bottom. The container is filled with water to a depth of 25 ± 5 mm above the bottom so that the prisms are approximately 10 mm clear of the water. A wick of absorbent material, e.g. terry cloth, is placed around the wall of the container so that the wick extends into the water (see Annex A4.7).

The containers shall be stored in a room or cabinet capable of being maintained at a temperature of 38 ± 2 °C throughout its storage space.

Note 1: *A suitable design of container is shown in Fig.* 1. *Further information on the containers is given in Annex* A4.7. *Alternatively, a reactor box can be used for storage (see Annex* B *in AAR-0).*

- Nominal capacity: 20 litres
- Dimensions: approx. 375 mm height x 315 mm diameter
- Easy opening and resealing airtight lid
- Suitable for dry or liquid applications
- Optional carrying handle
- Available in white, black or colours

Fig. 1 Storage containers suitable for the AAR-3 Concrete Prism Test

4.8 Casting and Curing Environment

For casting, a room maintained at 20 ± 2 °C (see also Annex A4.8).

The specimens shall be cured in a moist environment held at 20 ± 2 °C and not less than 90 % relative humidity. Initial measurements, immediately after demoulding, shall be made in a room maintained at 20 ± 2 °C and relative humidity not less than 50 % (see also Annex A4.8).

4.9 Measurement Environment

Prior to each measurement at 2, 4, 13, 26 and 52 weeks (see Sect. 8), the specimens are cooled in the sealed containers for 24 ± 2 h in a room at 20 °C, each measurement being made in a room maintained at 20 ± 2 °C and relative humidity not less than 50 % (see also Annex A4.8 for tropical countries).

Application 1—AAR-3.1: Method for Testing Potential Alkali-Reactivity of an Aggregate Combination

5 Materials

5.1 Sodium Hydroxide

Sodium hydroxide of at least technical grade (not less than 98 % sodium hydroxide by mass).

Note 2: *Caution—care should be taken when handling this chemical (and concrete containing it) and suitable protective clothing should be worn.*

5.2 Portland Cement

A Portland cement (EN197-1 [8] CEM I or ASTM C150 [9] Type I or similar) with total alkali content of 0.9–1.3 % sodium oxide equivalent (see Note 3). If necessary, sodium hydroxide shall be added to the concrete mix water so as to increase the alkali content of the binder to 1.25 ± 0.05 % sodium oxide equivalent and the alkali content of the mix is 5.5 ± 0.2 kg sodium oxide equivalent per cubic metre (see Note 3 and Annex A5.2).

Note 3: *Na_2O equivalent is defined as (% Na_2O + 0.658 % K_2O). There is evidence that adding NaOH does not always have the same effect on the concrete pore solution as using a cement manufactured with an equivalent total alkali content. Consequently, the use of a cement with the required high alkali content is preferred to the addition of NaOH. A suitable reference cement is described in AAR-0.*

5.3 Aggregates

5.3.1 Standard Aggregate Materials Test

The aggregate combination shall consist of one or more of the following:

(i) fine and coarse test aggregates;
(ii) fine test aggregate combined with a non-reactive coarse aggregate;
(iii) coarse test aggregate combined with a non-reactive fine aggregate.

The non-reactive fine or coarse aggregate shall have expansions in the AAR-2 mortar-bar test of less than 0.05 % at 14 days (Note 5).

Aggregate proportions for Standard Test: When planning the mix design for the standard test, the aggregates shall be combined in the following proportions (calculated on a saturated surface dry basis):

Fine aggregate—40 % (0–4 mm)

Coarse aggregate—60 % (4–22.4 mm)

The total aggregate grading should conform as far as possible to the recommended aggregate grading curve range in Table A.1 given in Annex A5.3. This is to ensure a concrete matrix suited for production of workable and stable concrete.

Note 4: *The coarse and fine aggregates should be 'clean'* (*i.e. a controlled content of fines, for example see criteria in EN 12620* [10] *or ASTM C33* [11]).

Note 5: *This expansion limit is deliberately more restrictive than is generally applied to identify non-reactive aggregate to ensure that the non-reactive aggregate has minimal effect on the results of this test.*

5.3.2 Alternative Test with Variable Aggregate Proportions

If it is desired to maximise the proportion of potentially reactive material in the test concrete, the highest amount of the appropriate aggregate fraction should be chosen from Table A.1, for example, 45 % of the total aggregate for a potentially reactive 0–2 mm sand, 55 % for a potentially reactive 0–4 mm sand or 65 % for a potentially reactive coarse aggregate. If there is doubt with respect to what fraction is to be considered as the most potentially reactive, it is recommended to test the possible “worst case” combinations of the fractions in question.

Note 6: *If a more precise investigation of the pessimum behaviour is desired, other combinations of the reactive and non-reactive aggregates can additionally be tested (see Annex* A5.3 *Note A17).*

6 Concrete Mix Design

(See also Annex A6).

The concrete mix design is given in Table 1.

Note 7: *If the requirement for maximum air content is difficult to fulfil, a defoaming agent should be applied. However, this must not contribute significant alkalis to the concrete. If it does, this must be included in the calculation of the added sodium hydroxide (see Annex* A5.2*).*

The mixing procedure is:

First pre-wet the fine aggregate for a period of at least 16 h to a total water content of 5 ± 2 %.

Then proceed as in the sequence shown in Table 2.

Before mixing the concrete, the mixer should be wiped out with wet cotton cloth in order to humidify the mixer.

Table 1 General concrete mix design

Cement content	440 kg/m^3
Free (effective) water content	220 kg/m^3
Coarse aggregate	60 % by mass (or as specified in Sect. 5.3.2)
Fine aggregate	40 % by mass (or as specified in Sect. 5.3.2)
Air content	<3 %

Table 2 Mixing sequence and periods

Add materials in sequence	Add materials (seconds from start of mixing)	Total mixing time (s)
Sand + Coarse aggregate	0	60
½ water + added NaOH (see Annex A5.2)	60	120
Pause	120	180
Cement + ½ water (+plasticizer if needed)	180	300

Note 8: *Worked examples of mix design calculation are given in Annex* A6.

Note 9: *The free (or effective) water is the water available for hydration of cement and for the workability of the fresh concrete. The total water added to the mix is the free water plus water absorbed by the aggregate to bring it to a saturated surface dry condition.*

Note 10: *It is important that the mix is sufficiently workable (ideally having a slump in the range 100–180 mm) to enable good compaction. With some flaky aggregates the above mix might not be sufficiently workable. If the slump of the concrete is too low, a superplasticizer with a low alkali content (not one combined with an air-entraining agent) should be used. In this case, the (small) amount of alkali contributed by the superplasticizer should be taken into account in calculating the amount of additional sodium hydroxide to be added (see Annex* A5.2*). The free water in the amount of superplasticizer expected to be added should also be taken into account in calculating the total content of water to be added (see Annex* A6*). The slump should not exceed 180 mm (even when using some superplasticizer).*

7 Test Specimens

If using mould oil to assist the release of the cast specimens, avoid the use of too much oil and ensure the reference studs are kept free of the oil.

Cast three test prisms from each concrete mix to be investigated, compacting the concrete into the moulds in two layers of equal depth using mechanical vibration. Compact sufficiently, but ensure no segregation and avoid as far as possible smoothing the prisms after vibration.

Cure at 20 ± 2 °C in relative humidity of not less than 90 % under moist covers for 24 ± 1 h (see Annex A4.8).

Record the weights of the covered moulds directly after casting and before demoulding to ensure there has been minimal water loss during the first 24 h curing.

8 Storage and Measurement

8.1 Initial Measurement Procedure

Before demoulding, identify and indelibly mark each concrete prism. After demoulding, identify and indelibly mark the top and the bottom of each concrete prism. Then examine it and record any defects.

Note 11: *De-mould and measure the prisms in sets of three. Do not de-mould all the prisms and attempt to measure consecutively as this will result in too much drying which could affect the expansive behaviour.*

Immediately weigh the prisms to the nearest ±0.1 g (W_0) and measure the initial length (*l*) of each prism to the nearest 1 mm using a steel rule. Clean the reference studs and take an initial reading on the comparator (C_o) using the Invar® rod to calibrate the length of the measuring apparatus. For each measurement, keep the prism in the same position (top and bottom in position, the same prism face towards the operator). Ensure the prism is well located in the measuring apparatus before making any readings. These and all subsequent measurements are to be made at a temperature of 20 ± 2 °C (see also Annex A4.8).

Verify that there is at least 20 mm of water in the container and place the prisms inside. Replace and seal the lid and store the container at 38 ± 2 °C until required for further measurement. 24 ± 2 h before making each set of further measurements, remove the container containing the prisms to be measured from the cabinet or room and allow the prisms to cool at 20 ± 2 °C whilst the prisms remain inside the container.

8.2 Procedures for Length and Weight Measurements

Remove the container from 38 ± 2 °C storage. DO NOT OPEN the container. Store the container at 20 ± 2 °C for 24 ± 2 h.

Make an initial reading on the comparator using the Invar® rod to calibrate the length of the measurement apparatus.

Each prism measurement should take no more than 2 min to avoid excessive drying. Follow the steps 1–5, to measure the prisms in each container:

1. Take a prism from the container, replacing the container cover immediately;
2. Remove any excess of moisture and weigh the prism (W_t where t = time in weeks), to the nearest ±0.1 g;
3. Clean the reference studs carefully;

4. Measure the prism using the comparator (C_t). For each measurement, keep the prism in the same position (top and bottom in position, the same prism face towards the operator). Ensure the prism is well located in the measuring apparatus before making any readings. After each measurement examine the prisms and note and report any cracking, gel exudations, warping or other features;
5. Return the prism to the container, upside down, so that the prisms are inverted after each measurement; then
6. Once all the prisms from one container have been measured, check the reading of the Invar® rod. If the difference is more than 3 μm from the first measurement, re-measure the prisms for this container again, applying steps 1–5;
7. Check the water level in the container (25 ± 5 mm), add water if necessary and replace the container cover.

Repeat steps 1–7 for all containers.

8.3 Measurement Timetable

After demoulding (see Sect. 8.1) and at the end of periods 2, 4, 13, 26 and 52 weeks after mixing (Note 12), measure the prisms (W_t and C_t). Do not remove the prisms from the container before the first 2 weeks of testing is completed. Do not make intermediate measurements.

Note 12: *Some types of slowly reacting aggregate may not exhibit deleterious expansion after 52 weeks. Depending on local experience, the duration of the test may be extended.*

9 Expression and Reporting Results

9.1 Calculations

Calculate the increase in length and weight for each prism for each period of measurement from the difference between the initial comparator or weight measurement (C_o, W_0) and the comparator or weight measurement after that period (C_t, W_t). Calculate each length increase as a percentage of the initial length of the corresponding prism to the nearest 0.001 % and the weight increase as a percentage of the initial weight of the corresponding prism to the nearest 0.01 %. Calculate the percentage net weight change for each prism from the beginning (W_0) to the end of the measurement (W_{52}).

For example, at 52 weeks the percentage length change E_{52} of a prism is given by:

$$E_{52} = \frac{C_{52} - C_0}{l} \times 100$$

where C_{52} is the comparator measurement at 52 weeks age, C_o is the initial comparator measurement of the prism, and l is the initial length of that prism (in mm).

For each measurement age, also calculate the mean length change of the three prisms to the nearest 0.005 % and the mean weight change to the nearest 0.01 %.

For example, at 52 weeks the mean percentage length change mE_{52} of the three test prisms is given by:

$$mE_{52} = \frac{E_{52/1} + E_{52/2} + E_{52/3}}{3}$$

9.2 Reporting

Report each length increase as a percentage of the initial length for the corresponding prism to the nearest 0.001 % and the weight increase as a percentage of the initial weight of the corresponding prism to the nearest 0.01 %.

Measurement of any weight loss (X_t) confirms there is insufficient water present in the system. AAR reactive mixes will not necessarily exhibit expansion if insufficient water is available. If a net weight loss is recorded at the time of executing the last length readings, the measurements relating to these prisms shall be discarded.

Note 13: *Re-wetting would not necessarily allow any previous expansive reactions to be reactivated and subsequent results would be unreliable. Experience has shown that weight gain in the test is related to curing conditions; especially, high humidity early curing can later lead to relatively low weight gain in the test, but results associated with unexpectedly low weight gains or even small weight losses should still be treated with caution.*

The measurements shall also be discarded if, at any time during the test period, either no water is left in the bottom of the storage container or the water in the bottom of the storage container has increased and reached the prisms allowing them to suck water (in which case the extent of leaching of alkalis from the prisms will be high).

For each measurement age, report the mean length change of the three prisms to the nearest 0.005 % and the mean weight change to the nearest 0.01 %.

Additionally report:

- the mix design;
- the prism size used;
- the cement source and alkali content;
- any alkali additions;
- any superplasticizer additions and any alkali contributed by the superplasticizer;
- the slump of concrete used in the prisms;
- observations regarding any cracking, gel exudations, warping or other features.

Note 14: *If the prisms exhibit an average expansion which is judged to be deleterious, it is recommended that a microscopical examination of the interior of the prism which has expanded the most should be carried out, together with an examination of the internal features and crack pattern, to confirm that the expansion is caused by alkali-silica reaction.*

Application 2—AAR-3.2: Method for Determining the Critical Alkali Threshold for an Aggregate Combination

Use materials according to 5 and the concrete mix according to 6.

Prepare at least four concrete mixes with alkali levels in increments normally between 2 and 5 kg/m^3 sodium oxide equivalent. Where possible, it is preferable to blend different Portland cements to achieve the desired alkali contents. If it is necessary to add sodium hydroxide, see examples in A5.2. for calculation of the required amount.

Add the sodium hydroxide, if necessary, to the mixing water before mixing (see Table 2), ensuring complete dissolution before use.

Cast and store the specimens according to 7 and 8 and report the results according to 9. The test period should usually be 52 weeks, but if it is known that the aggregate is of a slow reacting type, the test duration may need to be extended. For example many greywackes react at relatively low alkali levels and may need to be defined as high reactivity, but they need to be tested over a period of 2 years.

The alkali threshold is the highest alkali increment that is found not to give a significant expansion according to the criteria given in AAR-0. The use of the alkali threshold to designate the alkali reactivity is described in AAR 7.1, where the application of a safety margin to the result is explained.

Annex A—Additional Information

(Comments relate to clauses as numbered in the method)

A4.7 Specimen Storage

The seal of the airtight lid is critical in preventing significant loss of the water in the pail and hence in maintaining high humidity around the specimens. It is important regularly to check the water level in the container and to refill as necessary.

Plastic racks can be used to hold the prisms in place vertically in the container. The racks with the prisms are placed in the containers which are lined with a layer of either terry cloth or geotextile, ~7 mm in thickness.

The plastic racks should be made in such a way that there is enough clearance at the bottom to keep the prisms about 40–50 mm above the bottom of the pail. About 25 mm of water would then be placed at the bottom of the container.

The temperature-controlled room can be either different size ovens (which have a fairly limited capacity in some cases) or walk-in insulated chambers. Figures A.1 and A.2 show a typical laboratory set-up.

A4.8 Casting, Curing and Measuring Environment

Cooling to a standard temperature is necessary to obtain reproducible results. In countries with hot climates, however, it may occasionally be necessary to allow the casting, curing and measurement to be carried out in a room maintained at a temperature higher than the preferred 20 °C, up to a maximum of 27 ± 2 °C and not less that 65 % ± 5 RH.

Note A15: *A controlled temperature of 20 °C is preferred and strongly recommended for this stage of the test. Whatever standard temperature might be adopted (20 °C or another value up to 27 °C) for a particular test, it is essential that the same procedure and the same temperature are used consistently throughout the test. However, use of a temperature other than 20 °C will necessitate reconsideration of the expansion limits.*

A5.2 Calculation of Sodium Hydroxide to Be Added to the Mixing Water

Example calculation for determining the amount of sodium hydroxide (NaOH) to be added to the mixing water to increase the alkali content up to 1.25 % sodium oxide equivalent by mass of cement:

Fig. A.1 Concrete prism specimens in the storage container

Fig. A.2 Storage of containers

Cement content of 1 m^3 of concrete = 440 kg
Assumed content of superplasticizer = 2.0 kg
Specified amount of Na_2O eq. in concrete = 440 × 0.0125 = 5.50 kg
Amount of Na_2O eq. in cement (1.0 %) = 440 × 0.01 = 4.40 kg
Amount of Na_2O eq. in superplasticizer (1.0 %) = 2 × 0.01 = 0.02 kg
Amount of Na_2O eq. to be added per m^3 = 5.50 – 4.40 – 0.02 = 1.08 kg
The conversion factor Na_2O eq. to NaOH is 1.291.
Amount of NaOH required (to be added to the mixing water) = 1.08 × 1.291 = 1.39 kg
The purity of the technical grade NaOH to be used is 98 %
Amount of technical grade NaOH required (to be added and mixed together with the first half of the mixing water) = (1.39/98) × 100 = 1.42 kg

A5.3 Aggregates—Standard Aggregate Materials Test

The aggregate fractions should be combined in mass proportion calculated on a **saturated surface dry** basis according to Table A.1 given below which gives the recommended aggregate grading curve range to achieve a matrix suited for production of workable and stable concrete. In order to meet these requirements, it is necessary to know the particle size distribution curve of each aggregate.

Note A16: *Gaps between two fractions should be avoided. For example, sand 0/2 (2–0 mm) should not be tested with an aggregate 8/16 (16–8 mm).*

Note A17: *With some reactive aggregates it has been found that there is a proportion of reactive constituents in the aggregate that leads to a maximum expansion. This proportion is called the 'pessimum' content, and the relationship between expansion and content of reactive constituents is called the 'pessimum behaviour' of the reactive aggregate.*

Table A.1 Recommended aggregate grading curve range

Aperture size (mm)	Percentage by mass passing the test sieve[a]
22.4	100
16.0	75–95
8.0	45–70
4.0	35–55
2.0	25–45
1.0	20–35
0.5	10–25
0.25	4–12
0.125	1–8

[a]The range is selected to accommodate both crushed and uncrushed aggregate

A6 Concrete Mix Design

Preparation of the aggregates

Pre-wetted fine aggregates are used in order to ensure a workable and homogeneous concrete. This shall be done using the following procedure:

Determine the water absorption (WA_f) and initial water content (W_i) of the fine aggregate according to the standards/methods valid in the place of use. Weigh out a sufficient quantity of the fine aggregate, place it into a tray and add a sufficient amount of water calculated to achieve a water content of 5 ± 2 %. Mix the wetted fine aggregate thoroughly and store in sealed containers for at least 16 h. Measure the final water content (W_f) of the fine aggregate.

The coarse aggregate should be used without pre-wetting, but the water absorption (WA_c) and the water content (W_c) should be measured.

The amount of water added to the mix to achieve the prescribed free water content of 220 kg/m^3 is then calculated, taking into account both the water absorption and the measured water content of the fine and coarse aggregates. The water absorptions of most fine aggregates will be less than the intended water content (i.e. 5 ± 2 %). For most coarse aggregates, the water content will be lower than their water absorptions. In the cases where the water content (W) is higher than the water absorption (WA), the contribution from the aggregate (W_{free}) to the free water content of the mix can be calculated as:

Contribution from the aggregate = [(measured water content of the aggregate – water absorption of the aggregate)/100] × [calculated mass of the dry aggregate]

or

$$W_{free} = [(W - WA)/100] \times [(\text{quantity}_{ssd} \text{ of the aggregate})/(1 + WA/100)] \quad (A.1)$$

In the case of where the water absorption of the aggregate is higher than its water content, the extra water needed (W_{extra}) at the mixing stage, to bring the aggregate to a saturated surface dry state, can be calculated as:

Extra water needed = [(water absorption of the aggregate – measured water content of the aggregate)/100] × [calculated mass of the dry aggregate]

or

$$W_{extra} = [(WA - W)/100] \times [(\text{quantity}_{ssd} \text{ of the aggregate })/(1 + WA/100)] \quad (A.2)$$

Concrete mix design

A worked example is presented below, see Tables A.2, A.3, A.4, A.5 and A.6.

An example of concrete mix design calculation made with one coarse and one fine aggregate is shown (the calculation is the same for more aggregates).

Table A.2 General concrete mix design

Cement content	440 kg/m^3
Free (effective) water content	220 kg/m^3
Coarse aggregate	60 % by mass or as specified in Sect. 5.3.2
Fine aggregate	40 % by mass or as specified in Sect. 5.3.2
Air content	<3 %

Table A.3 Typical properties of the materials

	ρ_{rd} (kg/L)	ρ_{ssd} (kg/L)	Density (kg/L)	Water absorptionWA (%)	Water content W (%)
Coarse aggregate	2.66	2.68	–	0.8	0.3
Fine aggregate	2.65	2.68	–	1.1	6.2
Cement	–	–	3.08	–	–
Water	–	–	1.00	–	–
Sodium hydroxide	–	–	2.1	–	–
Superplasticizer	–	–	1.0–1.2	–	60–90

Table A.4 Calculation of the concrete mix design—stage 1

	Grading (mm)	Proportion[a] (%)	Quantity (kg/m^3)	Volume (L/m^3)*
Coarse aggregate	4/16	60	X	X/2.68
Fine aggregate	0/4	40	4/6 × X	(4/6 × X)/2.68
Cement	–	–	440	(440/3.08) = 143
Water (free)	–	–	220	220
Air (assumed 1.5 %)	–	–	–	15
Superplasticizer	–	–	2	~2
Sodium hydroxide	–	–	1.4	~1
Total	–	–	–	1000**

[a]According to Sect. 5.3.1
*Based on ρ_{ssd}—saturated surface dry aggregate
**The aim is to calculate the concrete mix design in order to make 1 m^3 of concrete

For the mix design it is necessary to know:

- The water absorption (WA) of each aggregate. The values can be determined according to EN 1097-6 [12], ASTM C127 [13] or ISO 7033 [14];
- The densities on an oven dry (ρ_{rd}) and saturated surface dry (ρ_{ssd}) basis of all fractions. The values can be determined according to EN 1097-6 [12], ASTM C127 [13] and 128 [15] or ISO 7033 [14];
- The water content (W) of each aggregate. The values can be determined according to EN 1097-5 [16] or ASTM C566 [17].

Table A.5 Calculation of the concrete mix design—stage 2

	Grading (mm)	Quantity (kg/m^3)	Volume (L/m^3)*
Coarse aggregate	4/16	995*	(995/2.68) = 371
Fine aggregate	0/4	663*	(663/2.68) = 247
Cement	–	440	143
Water (free)	–	220	220
Air	–	–	15
Superplasticizer	–	2	~2
Sodium hydroxide	–	1.4	~1
Total	–	2320	~1000

*Based on ρ_{ssd}—saturated surface dry aggregate

Table A.6 Final concrete mix design—quantities to be added to the mix

	Grading (mm)	Quantity (kg/m^3)	Quantity (kg/m^3)
Coarse aggregate	4/16	995*	995 – 5 = 990
Fine aggregate	0/4	663*	663 + 33 = 696
Cement	–	440	440
Water (added)	–	220	220 + 5 – 33 – 2 = 190
Air	–	–	–
Superplasticizer	–	2	2**
Sodium hydroxide	–	1.4	1.4
Total	–	2320	2320

*Based on ρ_{ssd}—saturated surface dry aggregate
**Free water content is 2 kg/m^3 × 80/100 = 1.6 kg/m^3 (~2 kg/m^3)

Assume that the content of superplasticizer (with density 1.1 kg/m^3 and water content 80 %) required is 2.0 kg/m^3 and that the amount of technical grade NaOH (with density 2.1 kg/m^3) required is 1.4 kg/m^3—see example in Annex A5.2.

The volumetric weight and the air content of the fresh concrete should be measured, according respectively to EN 12350-7 [18]/ASTM C1170 [19] and EN 12350-6 [20]/ASTM C173 [21]/ASTM C231 [22]. The ratio between the theoretical and the measured volumetric weights of the fresh concrete should be 1.000 ± 0.015 (which corresponds to 440 ± 6 kg of cement). This will ensure the correct amount of alkali from the cement in the fresh concrete.

Calculation of the quantity (X) of coarse aggregate (saturated surface dry) in 1 m^3 of concrete:

$$\text{Aggregate volume} = 1000 - 143 - 220 - 15 - 2 - 1 = 619\ L/m^3$$
$$(X/2.68\ kg/L) + (4/6 \times X)/2.68\ kg/L = 619\ L/m^3$$
$$X = 995\ kg/m^3$$

Calculation of the quantity (4/6 × X) of fine aggregate (saturated surface dry) in 1 m^3 of concrete:

$$4/6 \times X = 4/6 \times 995\ kg/m^3 = 663\ kg/m^3$$

The contribution from the fine aggregate ($W_{free\text{-}f}$) to the free water content can be calculated according to Eq. A.1:

$$W_{free\text{-}f} = [(6.2 - 1.1)/100] \times [(663)/(1 + 1.1/100)] = 33.4\ kg\ (\sim 33\ kg)$$

The extra water needed ($W_{extra\text{-}c}$) at the mixing stage to bring the coarse aggregate to a saturated surface dry state can be calculated according to Eq. A.2:

$$W_{extra\text{-}c} = [(0.8 - 0.3)/100] \times [(995)/(1 + 0.8/100)] = 5.0\ kg$$

Example of fresh concrete properties measured:

Slump = 120 mm (target is 100–180 mm)
Air content = 1.4 % (target <3 %)
Measured volumetric weight = 2327 kg/m^3

Ratio between the theoretical and measured volumetric weights = (2320/2327) = 0.997 (target is 0.985–1.015).

If the ratio is outside the range, the quantity of aggregate should be increased/decreased (without changing the proportion between the different aggregates) in order to meet the requirement of the weight ratio (0.985–1.015).

References

1. RILEM, 2000, Recommended test method TC 106-03, Detection of potentially alkali-reactivity of aggregates: B: Method for aggregate combinations using concrete prisms, *Materials and Structures*, 33 (229), 290–293.
2. Lindgård, J, Nixon, P J, Borchers, I, Schouenborg, B, Wigum, B J, Haugen M, Åkesson, U, The EU Partner project – European standard tests to prevent alkali reactions in aggregates, *Cement and Concrete Research*, 40 (2010) 611–635.
3. ASTM C1293-08b, Standard test method for determination of length change of concrete due to alkali-silica reaction.
4. ASTM C136, Standard test method for the sieve analysis of coarse and fine aggregates.
5. EN 480-1:2014, Admixtures for concrete, mortar and grout. Test methods. Reference concrete and reference mortar for testing.
6. EN 12390-2:2009, Testing hardened concrete. Making and curing specimens for strength tests.
7. ASTM C 192/C192M, Standard practice for making concrete test specimens in the laboratory.
8. EN 197-1:2011, Cement. Composition, specifications and conformity criteria for common cements.
9. ASTM C150-07, Standard specification for Portland cement.
10. EN 12620:2013, Aggregates for concrete.
11. ASTM C33/C33M-08, Standard specification for concrete aggregates.
12. EN 1097-6:2013, Tests for mechanical and physical properties of aggregates. Determination of particle density and water absorption.

13. ASTM C127-04, Standard test method for density, relative density (specific gravity), and absorption of coarse aggregate.
14. ISO 7033:1987 (R08), Fine and coarse aggregates for concrete - Determination of the particle mass-per-volume and water absorption - Pycnometer method.
15. ASTM C128-07, Standard test method for density, relative density (specific gravity), and absorption of fine aggregate.
16. EN 1097-5:2008, Tests for mechanical and physical properties of aggregates. Determination of water content by drying in a ventilated oven.
17. ASTM C566-97(2004), Standard test method for total evaporable moisture content of aggregate by drying.
18. EN 12350-7:2009, Testing fresh concrete. Air content. Pressure methods.
19. ASTM C1170-06, Standard test method for determining consistency and density of roller compacted concrete using a vibrating table.
20. EN 12350-6:2009, Testing fresh concrete. Density.
21. ASTM C173-01, Standard test method for air content of freshly mixed concrete by the volumetric method.
22. ASTM C231-04, Standard test method for air content of freshly mixed concrete by the pressure method.

RILEM Recommended Test Method: AAR-4.1—Detection of Potential Alkali-Reactivity—60 °C Test Method for Aggregate Combinations Using Concrete Prisms

1 Foreword

This is a method developed by RILEM in the light of the results of an international trial. This trial showed that the method can reliably differentiate reactive and non-reactive combinations for a range of aggregate compositions from around the world. It also takes into account the experience of the European PARTNER programme [1]. It is one of a suite of test methods for aggregate reactivity developed by RILEM, the combined use of which for aggregate assessment is given in AAR-0. This method is developed from an AFNOR performance test for concrete mixes [2–4].

2 Scope

The method covers the measurement of expansion produced by alkali-aggregate reaction of concrete prisms stored in an environment which accelerates the reaction. It enables the testing of the potential alkali-reactivity of aggregate combinations. The method has been developed for *normal weight* aggregates and is not necessarily applicable for lightweight (*oven dry particle density less than 2000* kg/m^3) or heavy weight aggregates (*oven dry particle density greater than 3500* kg/m^3).

Additional information concerning the testing methodology is given in Annex A.

3 Principle

Concrete test prisms are prepared from the aggregate combination under test (either the standard test with fixed coarse:fine aggregate proportions, or the alternative test with variable aggregate proportions) and are stored in hot (60 °C), humid conditions

P.J. Nixon and I. Sims (eds.), *RILEM Recommendations for the Prevention of Damage by Alkali-Aggregate Reactions in New Concrete Structures*,
RILEM State-of-the-Art Reports 17, DOI 10.1007/978-94-017-7252-5_6

for at least 20 weeks to promote any alkali-aggregate reaction. Measurements are made at periodic intervals to determine whether any expansion has occurred. To promote further any potential for reaction the prisms are made with a relatively high content of Portland cement with a high alkali content such that the alkali level in the concrete is 5.5 kg sodium oxide equivalent per cubic metre of concrete. Sodium hydroxide is added to the mix when necessary to enhance the alkali level.

In order to identify the effect of specific aggregate combinations and any pessimum effects, there are options to test the coarse and fine test aggregates together or either in combination with a non-reactive material.

4 Apparatus

4.1 Sieve

Aperture sizes (mm) 22.4, 16.0, 8.0, 4.0, 2.0, 1.0, 0.5, 0.25 and 0.125 or ASTM equivalents [5].

4.2 Moulds

Three steel moulds suitable for casting concrete prisms of lengths 250 ± 50 mm and cross-section 75 ± 5 mm. The moulds shall have the facility for casting stainless steel reference studs into the mid-points of the end faces of the prisms.

4.3 Length Comparator

The length comparator should be such as to accommodate the shape of the reference studs in the prism and shall incorporate a gauge mounted rigidly in a vertical orientation. The graduation of the gauge shall not be greater than 0.002 mm and the error throughout the range of traverse shall be no more than 0.005 mm. Horizontally mounted comparators are not suitable for this test method.

4.4 Standard Length Gauge

This shall consist of an Invar® rod with ends machined to accommodate the reference studs.

4.5 *Weighing Device*

Suitable for weighing specimens to the nearest 0.1 g.

4.6 *Concrete Mixing and Casting Equipment*

This shall be suitable for mixing batches (EN 480-1 [6]) and vibrating into the moulds (EN 12390-2 [7] or ASTM C 192/C192 M [8]).

4.7 *Specimen Storage*

The specimens are stored in containers within a reactor (Fig. 1) generating 60 ± 2 °C and a relative humidity as close as possible to 100 %. The water level in the bottom of the reactor is approximately 190 mm.

Sealable containers (Fig. 2) with 35 ± 5 mm water (see Note 2) in the bottom, with prisms located 15 mm above the water. The standard containers are designed to contain 3 specimens. Alternative containers suitable for 6 specimens can be used; appropriate dimensions would be 400 x 280 x 230 mm.

Note1: *Some suppliers of reactors are given in AAR-0.*

Note 2: *In (mainly research) cases where any alkali loss from the concrete prisms during testing is being monitored by periodically analysing the water in the bottom of the sealable containers, consideration should be given to the use of deionised water.*

4.8 *Casting and Curing Environment*

For casting, a room maintained at 20 ± 2 °C (see also Annex A4.8).

The specimens shall be cured in a moist environment held at 20 ± 2 °C and not less than 90 % relative humidity. Initial measurements, immediately after demoulding, shall be made in a room maintained at 20 ± 2 °C and relative humidity not less than 50 % (see also Annex A4.8).

4.8.1 Measurement Environment

Prior to each measurement at 5, 10, 15 and 20 weeks (see Sect. 8.2) the specimens are cooled in the sealed containers for 24 ± 2 h in a room at 20 ± 2 °C, each

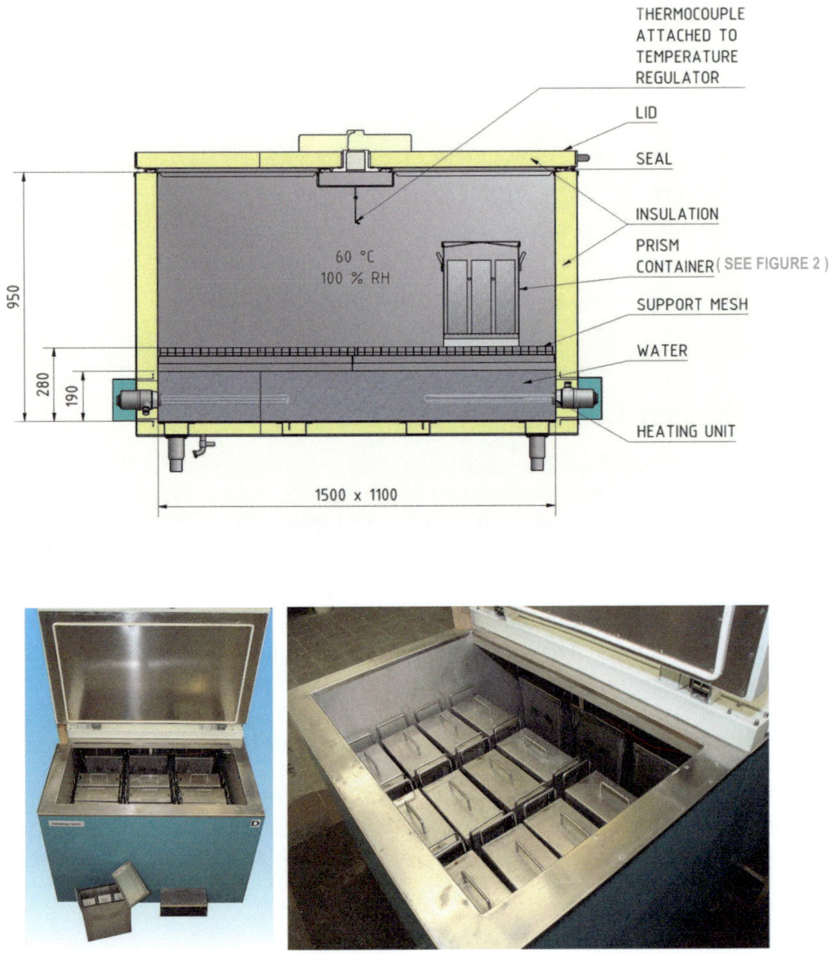

Fig. 1 Reactor generating 60 °C and 100 % Humidity (dimensions in mm)

measurement being made in a room maintained at 20 ± 2 °C and relative humidity not less than 50 % (see Annex A4.8).

5 Materials

5.1 Sodium Hydroxide

Sodium hydroxide of at least technical grade (not less than 98 % sodium hydroxide by mass).

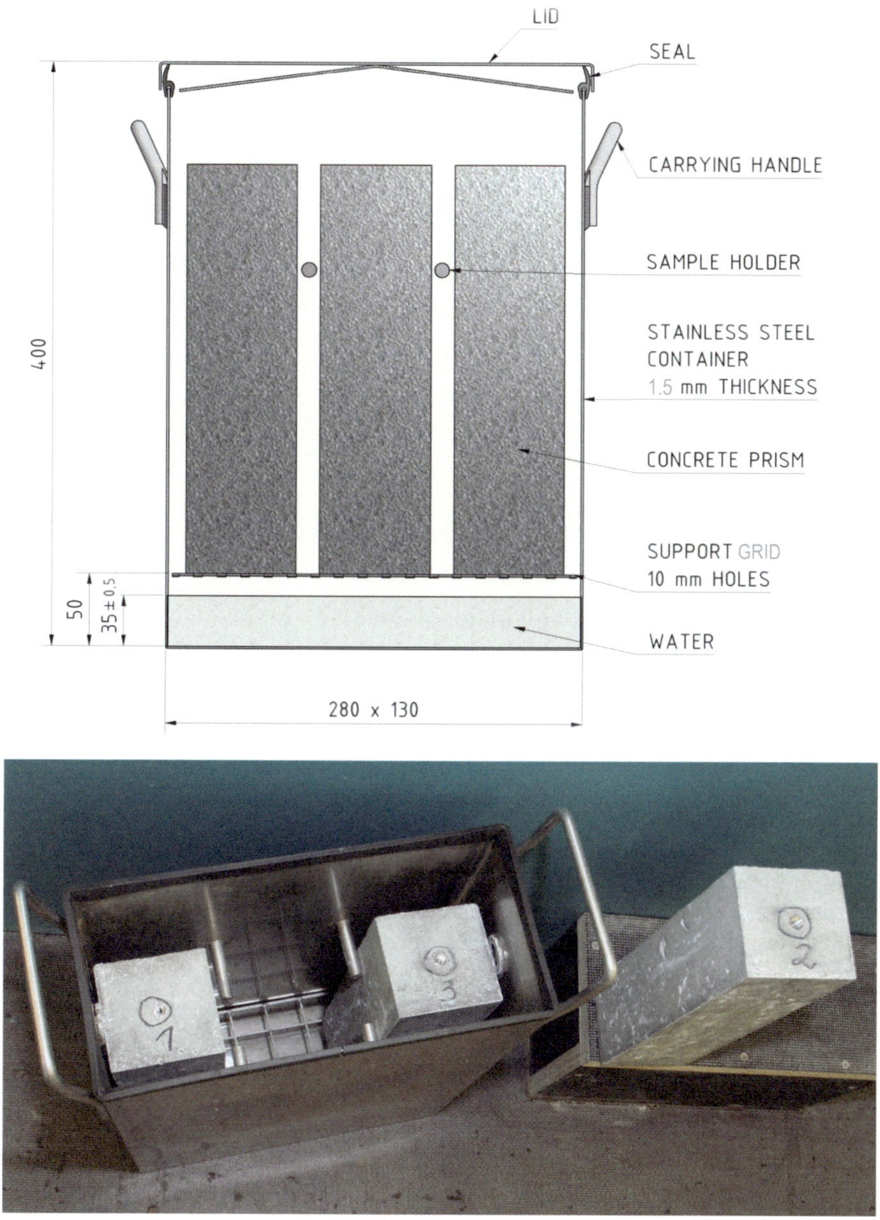

Fig. 2 Sealable containers for storing concrete prisms (dimensions in mm)

Note 3: *CAUTION—care should be taken when handling this chemical (and concrete containing it) and suitable protective clothing should be worn.*

5.2 *Portland Cement*

A Portland cement (EN197-1 [9] CEM I or ASTM C150 [10] Type I or similar) with total alkali content of 0.9–1.3 % sodium oxide equivalent (see Note 4). If necessary, sodium hydroxide shall be added to the concrete mix water so as to increase the alkali content of the binder to 1.25 ± 0.05 % sodium oxide equivalent and the alkali content of the mix to 5.5 ± 0.2 kg sodium oxide equivalent per cubic metre (see Note 4 and Annex A5.2).

Note 4: *Na_2O equivalent is defined as (% Na_2O + 0.658 % K_2O). There is evidence that adding NaOH does not always have the same effect on the concrete pore solution as using a cement manufactured with an equivalent total alkali content. Consequently, the use of a cement with the required high alkali content is preferred to the addition of NaOH. A suitable reference cement is described in AAR-0.*

5.3 *Aggregates*

5.3.1 Standard Aggregate Materials Test

The aggregate combination shall consist of one or more of the following:

(i) Fine and coarse test aggregates;
(ii) Fine test aggregate combined with a non-reactive coarse aggregate;
(iii) Coarse test aggregate combined with a non-reactive fine aggregate.

The non-reactive fine or coarse aggregate shall have expansions in the AAR-2 mortar-bar test of less than 0.05 % at 14 days (Note 6).

Aggregate proportions for Standard Test: when planning the mix design for the standard test the aggregates shall be combined in the following proportions (calculated on a saturated surface dry basis):

Fine aggregate—40 % (0–4 mm)

Coarse aggregate—60 % (4–22.4 mm)

The total aggregate grading should conform as far as possible to the recommended aggregate grading curve range in Table A.1 given in Annex A5.3. This is to enable a concrete matrix suited for production of workable and stable concrete.

Note 5: *The coarse and fine aggregates should be 'clean'* (*i.e. a controlled content of fines, for example see criteria in EN 12620* [11] *or ASTM C33* [12]).

Note 6: *This expansion limit is deliberately more restrictive than is generally applied to identify non-reactive aggregate to ensure that the non-reactive aggregate has minimal effect on the results of this test.*

5.3.2 Alternative Test with Variable Aggregate Proportions

If it is desired to maximise the proportion of potentially reactive material in the test concrete, the highest amount of the appropriate aggregate fraction should be chosen from Table A.1, for example, 45 % of the total aggregate for a potentially reactive 0–2 mm sand, 55 % for a potentially reactive 0–4 mm sand or 65 % for a potentially reactive coarse aggregate. If there is doubt as to which fraction is to be considered as the most potentially reactive, it is recommended that the possible "worst case" combinations of the fractions in question are tested.

Note 7: *If a more precise investigation of the pessimum behaviour is desired, other combinations of the reactive and non-reactive aggregates can additionally be tested (see* Annex A5.3, *Note A18).*

6 Concrete Mix Design

(See also Annex A6).

The concrete mix design is given in Table 1.

Note 8: *If the requirement for maximum air content is difficult to fulfil, a de-foaming agent should be used. However, this must not contribute significant alkalis to the concrete. If it does, this must be included in the calculation of the added sodium hydroxide (see* Annex A5.2).

The mixing procedure is:

First pre-wet the fine aggregate for a period of at least 16 h to a total water content of 5 ± 2 %.

Then proceed as in the sequence shown in Table 2.

Before mixing the concrete, the mixer should be wiped out with wet cotton cloth in order to humidify the mixer.

Note 9: *Worked examples of mix design calculation are given in* Annex A6.

Table 1 General concrete mix design

Cement content	440 kg/m^3
Free (effective) water content	220 kg/m^3
Coarse aggregate	60 % by mass (or as specified in Sect. 5.3.2)
Fine aggregate	40 % by mass (or as specified in Sect. 5.3.2)
Air content	<3 %

Table 2 Mixing sequence and periods

Add materials in sequence	Add materials (seconds from beginning of mixing)	Total mixing time (seconds)
Sand + Coarse aggregate	0	60
½ water + added NaOH (see Annex A5.2)	60	120
Pause	120	180
Cement + ½ water (+plasticizer if needed)	180	300

Note 10: *The free (or effective) water is the water available for hydration of cement and for the workability of the fresh concrete. The total water added to the mix is the free water plus water absorbed by the aggregate to bring it to a saturated surface dry condition.*

Note 11: *It is important that the mix is sufficiently workable (ideally having a slump in the range 100–180 mm) to enable good compaction. With some flaky aggregates the above mix might not be sufficiently workable. If the slump of the concrete is too low, a superplasticizer with a low alkali content (not one combined with an air-entraining agent) should be used. In this case, the (small) amount of alkali contributed by the superplasticizer should be taken into account in calculating the amount of additional sodium hydroxide to be added (see* Annex A5.2*). The free water in the amount of superplasticizer expected to be added should also be taken into account in calculating the total content of water to be added (see* Annex A6*). The slump should not exceed 180 mm (even when using some superplasticizer).*

7 Test Specimens

If using mould oil to assist the release of the cast specimens, avoid the use of too much oil and ensure the reference studs are kept free of the oil.

Cast three test prisms from each concrete mix to be investigated, compacting the concrete into the moulds in two layers of equal depth using mechanical vibration. Compact sufficiently, but ensure no segregation and avoid as far as possible smoothing the prisms after vibration.

Cure at 20 ± 2 °C in relative humidity of not less than 90 % under moist covers for 24 ± 1 h (see Annex A4.8).

Record the weights of the covered moulds directly after casting and before demoulding to ensure there has been minimal water loss during the first 24 h curing.

8 Storage and Measurements

8.1 Initial Measurement Procedure

Before demoulding, identify and indelibly mark each concrete prism. After de-moulding, identify and indelibly mark the top and the bottom of each concrete prism. Then examine it and record any defects.

Immediately weigh the prisms to the nearest ±0.1 g (W_0) and measure the initial length (*l*) of each prism to the nearest 1 mm using a steel rule. Clean the reference studs and take an initial reading on the comparator (C_0) using the Invar® rod to calibrate the length of the measuring apparatus. For each measurement, keep the prism in the same position (top and bottom in position, the same prism face towards the operator). Ensure the prism is well located in the measuring apparatus before making any readings. These and all subsequent measurements are to be made at a temperature of 20 ± 2 °C (see also Annex A4.8).

Place the prisms in containers with 35 ± 5 mm of water at the bottom (see Sect. 4.7). Each sealable container holds a maximum of either six prisms or three prisms (Fig. 1). It is important that a full set of specimens (i.e. either 3 or 6) are in the container. If fewer test specimens are needed, the number should be made up with "dummy" specimens. Put all the sealed containers in the 60 ± 2 °C reactor (Fig. 2) and ensure the reactor contains an appropriate level of water (see Sect. 4.7).

Note 12: *De-mould and measure the prisms in sets of three. Do not de-mould all the prisms and attempt to measure consecutively as this will result in too much drying which could affect the expansive behaviour.*

8.2 Procedures for Length and Weight Measurements

Remove the sealed containers from the reactor and ensure the cover is securely in place. DO NOT OPEN the containers. Store the containers at 20 ± 2 °C for 24 ± 2 h, to cool them, in a high relative humidity chamber if possible.

Make an initial reading on the comparator using the Invar® rod to calibrate the length of the measurement apparatus.

Each prism measurement should take no more than 2 min to avoid excessive drying. Follow the steps 1–5, to measure the prisms in each container:

1. Take a prism from the container, replacing the container cover immediately;
2. Remove any excess of moisture and weigh the prism (W_t where t = time in weeks), to the nearest ±0.1 g;

3. Clean the reference studs carefully;
4. Measure the prism using the comparator (C_t). For each measurement, keep the prism in the same position (top and bottom in position, the same prism face towards the operator). Ensure the prism is well located in the measuring apparatus before making any readings. After each measurement, examine the prisms and note and report any cracking, gel exudations, warping or other features;
5. Return the prism to the container, upside down, so that the prisms are inverted after each measurement;
6. Once all the prisms from one container have been measured, check the reading of the Invar® rod. If the difference is more than 3 μm from the first measurement, re-measure the prisms for this container again, applying steps 1–5;
7. Check the water level in the container (35 ± 5 mm), add water if necessary and replace the container cover.

Repeat steps 1–7 for all containers. Check the water level in the reactor and add water if necessary. Replace all the containers into the reactor.

If using containers for six prisms, it is good practice to avoid measuring only three prisms out of the six in one container on any one date. Nevertheless if this occurs, proceed as follows:

Inside the reactor, open the container (E) in which the prisms to be measured are stored. Remove the prisms to be measured placing them in a second container (R) that contains an appropriate quantity of water and has been stored in the reactor for at least 4 h.

Remove the second container (R) from the reactor and store at 20 ± 2 °C for 24 ± 2 h in a high relative humidity chamber if possible, then proceed as described in Sect. 8.2, steps 1–6.

After measurement, inside the reactor, remove the prisms from the second container (R) and return them to the container (E), upside down, so that the prisms are inverted after each measurement.

8.3 Measurement Timetable

After demoulding (see Sect. 8.1) and at the end of periods 5, 10, 15 and 20 weeks after mixing (Note 13), measure the prisms (W_t and C_t). Do not remove the prisms from the reactor before the first 5 weeks of testing is completed. Do not make intermediate measurements.

Note 13: *Some types of slowly reacting aggregate may not exhibit deleterious expansion after 20 weeks. Depending on local experience, the duration of the test may be extended.*

9 Expression and Reporting Results

9.1 Calculations

Calculate the increase in length and weight for each prism for each period of measurement from the difference between the initial comparator or weight measurement (C_0, W_0) and the comparator or weight measurement after that period (C_t, W_t). Calculate each length increase as a percentage of the initial length of the corresponding prism to the nearest 0.001 % and the weight increase as a percentage of the initial weight of the corresponding prism to the nearest 0.01 %.

For example, at 20 weeks the percentage length change E_{20} of a prism is given by:

$$E_{20} = \frac{C_{20} - C_0}{l} \times 100$$

where

C_{20} is the comparator measurement at 20 weeks age,

C_0 is the initial comparator measurement of the prism and l is the initial length of that prism (in mm).

And the weight variation (X_{20}) is given by:

$$X_{20} = \frac{W_{20} - W_0}{W_0} \times 100$$

For each measurement age, also calculate the mean length change of the three prisms to the nearest 0.005 % and the mean weight change to the nearest 0.01 %.

For example, at 20 weeks the mean percentage length change mE_{20} of the three test prisms is given by:

$$mE_{20} = \frac{E_{20/1} + E_{20/2} + E_{20/3}}{3}$$

9.2 Reporting

Report each length increase as a percentage of the initial length for the corresponding prism to the nearest 0.001 % and the weight increase as a percentage of the initial weight of the corresponding prism to the nearest 0.01 %.

Measurement of any weight loss (X_t) confirms there is insufficient water present in the system. AAR reactive mixes will not necessarily exhibit expansion if insufficient water is available. If a net weight loss is recorded at the time of executing the last length readings, the measurements relating to these prisms shall be discarded.

Note 14: *Re-wetting would not necessarily allow any previous expansive reactions to be reactivated and subsequent results would be unreliable. Experience has shown that weight gain in the test is related to curing conditions; especially, high humidity early curing can later lead to relatively low weight gain in the test, but results associated with unexpectedly low weight gains or even small weight losses should still be treated with caution.*

The measurements shall also be discarded if, at any time during the test period, either no water is left in the bottom of the storage container or the water in the bottom of the storage container has increased and reached the prisms allowing them to suck water (in which case the extent of leaching of alkalis from the prisms will be high).

Calculate and report the mean length change of the three prisms for each measurement age to the nearest 0.005 % and the mean weight change to the nearest 0.01 %.

After each measurement examine the prisms and note and report any cracking, gel exudations, warping or other features.

Additionally report:

- the mix design;
- the prism size used;
- the cement source and alkali content;
- any alkali additions;
- any superplasticizer additions and any alkali contributed by the superplasticizer;
- the slump of concrete used in the prisms;
- the storage conditions (incl. type of reactor, type of containers and number of prisms stored in each container).

Note 15: *If the prisms exhibit an average expansion which is judged to be deleterious, it is recommended that a microscopic examination of the interior of the prism which has expanded the most should be carried out, together with an examination of the internal crack pattern, to confirm that the expansion is caused by alkali-silica reaction.*

Annex A—Additional Information

(Comments relate to clauses as numbered in the method)

A4.8 Casting and Measurement Environment

Cooling to a standard temperature is necessary to obtain reproducible results. In countries with hot climates, however, it may occasionally be necessary to allow the casting and measurement to be carried out in a room maintained at a temperature higher than the preferred 20 °C, up to a maximum of 27 ± 2 °C and not less that 65 ± 5 % RH.

Note A16: *A controlled temperature of 20 °C is preferred and strongly recommended for this stage of the test. Whatever standard temperature might be adopted (20 °C or another value up to 27 °C) for a particular test, it is essential that the same procedure and the same temperature are used consistently throughout the test. However, use of a temperature other than 20 °C will necessitate reconsideration of the expansion limits.*

A5.2 Calculation of Sodium Hydroxide to Be Added to the Mixing Water

Example calculation to determine the amount of sodium hydroxide (NaOH) to be added to the mixing water to increase the alkali content up to 1.25 % sodium oxide equivalent by mass of cement:

Cement content of 1 m^3 of concrete = 440 kg

Assumed content of superplasticizer = 2.0 kg

Specified amount of Na_2O eq. in concrete = 440 × 0.0125 = 5.50 kg

Amount of Na_2O eq. in cement (1.0 %) = 440 × 0.01 = 4.40 kg

Amount of Na_2O eq. in superplasticizer (1.0 %) = 2 × 0.01 = 0.02 kg

Amount of Na_2O eq. to be added per/m^3 = 5.50 − 4.40 − 0.02 = 1.08 kg

The conversion factor Na_2O eq. to NaOH is 1.291. Amount of NaOH required (to be added to the mixing water) = 1.08 × 1.291 = 1.39 kg

The purity of the technical grade NaOH to be used is 98 %. Amount of technical grade NaOH required (to be added and mixed together with the first half of the mixing water) = (1.39/98) × 100 = 1.42 kg

A5.3 Aggregate—Standard Aggregate Materials Test

The aggregate fractions should be combined in mass proportions calculated on a **saturated surface dry** basis according to Table A.1 given below, which gives the recommended aggregate grading curve range to achieve a matrix suited for production of workable and stable concrete. In order to meet these requirements, it is necessary to know the particle size distribution curve of each aggregate.

Table A.1 Recommended aggregate grading curve range

Aperture size (mm)	Percentage by mass passing the test sieve[a]
22.4	100
16.0	75–95
8.0	45–70
4.0	35–55
2.0	25–45
1.0	20–35
0.5	10–25
0.25	4–12
0.125	1–8

[a]The range is selected to accommodate both crushed and uncrushed aggregate

Note A17: *Gaps between two fractions should be avoided. For example, sand 0/2 (2–0 mm) should not be tested with an aggregate 8/16 (16–8* mm*).*

Note A18: *With some reactive aggregates it has been found that there is a proportion of reactive constituents in the aggregate that leads to a maximum expansion. This proportion is called the 'pessimum' content, and the relationship between expansion and content of reactive constituents is called the 'pessimum behaviour' of the reactive aggregate.*

A6 Concrete Mix Design

Preparation of the aggregates

Pre-wetted fine aggregates are used in order to ensure a workable and homogeneous concrete. This shall be done using the following procedure:

Determine the water absorption (WA_f) and initial water content (W_i) of the fine aggregate according to the standards/methods valid in the place of use. Weigh out a sufficient quantity of the fine aggregate, place it into a tray and add a sufficient amount of water calculated to achieve a water content of 5 ± 2 %. Mix the wetted fine aggregate thoroughly and store in sealed containers for at least 16 h. Measure the final water content (W_f) of the fine aggregate.

The coarse aggregate should be used without pre-wetting, but the water absorption (WA_c) and the water content (W_c) should be measured.

The amount of water added to the mix to achieve the prescribed free water content of 220 kg/m^3 is then calculated, taking into account both the water absorption and the measured water content of the fine and coarse aggregates. The water absorptions of most fine aggregates will be less than the intended water content (i.e. 5 ± 2 %). For most coarse aggregates, the water content will be lower than their water absorptions. In the cases where the water content (W) is higher than the water absorption (WA), the contribution from the aggregate (W_{free}) to the free water content of the mix can be calculated as:

$$\text{Contribution from the aggregate} = [(\text{measured water content of the aggregate} - \text{water absorption of the aggregate})]/100 \times [\text{calculated mass of the dry aggregate}]$$

or

$$W_{free} = [(W - WA)/100] \times [(\text{quantity}_{ssd}\text{ of the aggregate})/(1 + WA/100)] \quad (A.1)$$

In the case of where the water absorption of the aggregate is higher than its water content, the extra water needed (W_{extra}) at the mixing stage, to bring the aggregate to a saturated surface dry state, can be calculated as:

$$\text{Extra water needed} = [(\text{water absorption of the aggregate} - \text{measured water content of the aggregate})]/100 \times [\text{calculated mass of the dry aggregate}]$$

or

$$W_{extra} = [(WA - W)/100] \times [(\text{quantity}_{ssd}\text{ of the aggregate})/(1 + WA/100)] \quad (A.2)$$

Concrete Mix Design

A worked example is presented below, see Tables A.2, A.3, A.4, A.5 and A.6.

Example of concrete mix design calculation made with one coarse and one fine aggregate is shown (the calculation is the same for more aggregates).

For the mix design it is necessary to know:

- The water absorption (WA) of each aggregate. The values can be determined according to EN 1097-6 [13], ASTM C127 [14] or ISO 7033 [15];
- The densities on an oven dry (ρ_{rd}) and saturated surface dry (ρ_{ssd}) basis of all fractions. The values can be determined according to EN 1097-6 [13], ASTM C127 [14] and 128 [16] or ISO 7033 [15];
- The water content (W) of each aggregate. The values can be determined according to EN 1097-5 [17] or ASTM C566 [18].

Assume that the content of superplasticizer (with density 1.1 kg/m^3 and water content 80 %) required is 2.0 kg/m^3 and that the amount of technical grade NaOH (with density 2.1 kg/m^3) required is 1.4 kg/m^3—see example in Annex A5.2.

The volumetric weight and the air content of the fresh concrete should be measured, according respectively to EN 12350-7 [19]/ASTM C1170 [20] and EN 12350-6 [21]/ASTM C173 [22]/ASTM C231 [23]. The ratio between the theoretical and the measured volumetric weights of the fresh concrete should be

Table A.2 General concrete mix design

Cement content	440 kg/m^3
Free (effective) water content	220 kg/m^3
Coarse aggregate	60 % by mass or as specified in Sect. 5.3.2
Fine aggregate	40 % by mass or as specified in Sect. 5.3.2
Air content	<3 %

Table A.3 Typical properties of the materials

	ρ_{rd} (kg/L)	ρ_{ssd} (kg/L)	Density (kg/L)	Water absorption WA (%)	Water content W (%)
Coarse aggregate	2.66	2.68	–	0.8	0.3
Fine aggregate	2.65	2.68	–	1.1	6.2
Cement	–	–	3.08	–	–
Water	–	–	1.00	–	–
Sodium hydroxide	–	–	2.1	–	–
Superplasticizer	–	–	1.0–1.2	–	60–90

Table A.4 Calculation of the concrete mix design—stage 1

	Grading (mm)	Proportion[a] (%)	Quantity (kg/m^3)	Volume (L/m^3)*
Coarse aggregate	4/16	60	X	X/2.68
Fine aggregate	0/4	40	4/6 × X	(4/6 × X)/2.68
Cement	–	–	440	(440/3.08) = 143
Water (free)	–	–	220	220
Air (assumed 1.5 %)	–	–	–	15
Superplasticizer	–	–	2	−2
Sodium hydroxide	–	–	1.4	−1
Total	–	–	–	1000**

[a]According to Sect. 5.3.1 *Based on ρ_{ssd}—saturated surface dry aggregate
**The aim is to calculate the concrete mix design in order to make 1 m^3 of concrete

Table A.5 Calculation of the concrete mix design—stage 2

	Grading (mm)	Quantity (kg/m^3)	Volume (L/m^3)*
Coarse aggregate	4/16	995*	(995/2.68) = 371
Fine aggregate	0/4	663*	(663/2.68) = 247
Cement	–	440	143
Water (free)	–	220	220
Air	–	–	15
Superplasticizer	–	2	−2
Sodium hydroxide	–	1.4	−1
Total	–	2320	−1000

*Based on ρ_{ssd}—saturated surface dry aggregate

Table A.6 Final concrete mix design—quantities to be added to the mix

	Grading (mm)	Quantity (kg/m^3)	Quantity (kg/m^3)
Coarse aggregate	4/16	995*	995 − 5 = 990
Fine aggregate	0/4	663*	663 + 33 = 696
Cement	–	440	440
Water (added)	–	220	220 + 5 − 33 − 2 = 190
Air	–	–	–
Superplasticizer	–	2	2**
Sodium hydroxide	–	1.4	1.4
Total	–	2320	2320

*Based on ρ_{ssd}—saturated surface dry aggregate
**Free water content is 2 kg/m^3 × 80/100 = 1.6 kg/m^3 (−2 kg/m^3)

1.000 ± 0.015 (which corresponds to 440 ± 6 kg of cement). This will ensure the correct amount of alkali from the cement in the fresh concrete.

Calculation of the quantity (X) of coarse aggregate (saturated surface dry) in 1 m^3 of concrete:

$$\text{Aggregate volume} = 1000 - 143 - 220 - 15 - 2 - 1 = 619\ \text{L/m}^3$$
$$(X/2.68\ \text{kg/L}) + (4/6 \times X)/2.68\ \text{kg/L} = 619\ \text{L/m}^3$$
$$X = 995\ \text{kg/m}^3$$

Calculation of the quantity (4/6 × X) of fine aggregate (saturated surface dry) in 1 m^3 of concrete:

$4/6 \times X = 4/6 \times 995\ \text{kg/m}^3 = 663\ \text{kg/m}^3$

The contribution from the fine aggregate ($W_{free\text{-}f}$) to the free water content can be calculated according to Eq. A.1:

$$W_{free-f} = [(6.2 - 1.1)/100] \times [(663)/(1 + 1.1/100)] = 33.4\ \text{kg}\ (\sim 33\ \text{kg})$$

The extra water needed ($W_{extra\text{-}c}$) at the mixing stage to bring the coarse aggregate to a saturated surface dry state can be calculated according to Eq. A.2:

$$W_{extra-c} = [(0.8 - 0.3)/100] \times [(995)/(1 + 0.8/100)] = 5.0\ \text{kg}$$

Example of fresh concrete properties measured:

Slump = 120 mm (target is 100–180 mm)

Air content = 1.4 % (target <3 %)

Measured volumetric weight = 2327 kg/m^3

Ratio between the theoretical and measured volumetric weights = (2320/2327) = 0.997 (target is 0.985–1.015).

If the ratio is outside the range, the quantity of aggregate should be increased/decreased (without changing the proportion between the different aggregates) in order to meet the requirement of the weight ratio (0.985–1.015).

References

1. Lindgård, J, Nixon, P J, Borchers, I, Schouenborg, B, Wigum, B J, Haugen M, Åkesson, U, The EU *Partner* project – European standard tests to prevent alkali reactions in aggregates, *Cement and Concrete Research*, 40 (2010) 611–635.
2. LCPC, Recommendations for the prevention of damage by the alkali-aggregate reaction, Paris, France, 1994.
3. AFNOR NF P 18–454, Reactivity of a concrete formula with regard to the alkali-aggregate reaction, *Association Française de Normalisation*, Paris, France, December 2004.
4. AFNOR NF P 18-456, Reactivity of a concrete formula with regard to the alkali-aggregate reaction - Criteria for interpretation of the performance test results, *Association Française de Normalisation*, Paris, France November 2004.
5. ASTM C136-06, Standard test method for the sieve analysis of coarse and fine aggregates.
6. EN 480-1:2014, Admixtures for concrete, mortar and grout. Test methods. Reference concrete and reference mortar for testing.
7. EN 12390-2:2009, Testing hardened concrete. Making and curing specimens for strength tests.
8. ASTM C192/C192 M-07, Standard practice for making concrete test specimens in the laboratory.
9. EN 197-1:2011, Cement. Composition, specifications and conformity criteria for common cements.
10. ASTM C150-07, Standard specification for Portland cement.
11. EN 12620:2013, Aggregates for concrete.
12. ASTM C33/C33 M-08, Standard specification for concrete aggregates.
13. EN 1097-6:2013, Tests for mechanical and physical properties of aggregates. Determination of particle density and water absorption.
14. ASTM C127-04, Standard test method for density, relative density (specific gravity), and absorption of coarse aggregate.
15. ISO 7033:1987 (R08), Fine and coarse aggregates for concrete - Determination of the particle mass-per-volume and water absorption - Pycnometer method.
16. ASTM C128-07, Standard test method for density, relative density (specific gravity), and absorption of fine aggregate.
17. EN 1097-5:2008, Tests for mechanical and physical properties of aggregates. Determination of water content by drying in a ventilated oven.
18. ASTM C566-97(2004), Standard test method for total evaporable moisture content of aggregate by drying.
19. EN 12350-7:2009, Testing fresh concrete. Air content. Pressure methods.
20. ASTM C1170-06, Standard test method for determining consistency and density of roller compacted concrete using a vibrating table.
21. EN 12350-6: 2009, Testing fresh concrete. Density.
22. ASTM C173-01, Standard test method for air content of freshly mixed concrete by the volumetric method.
23. ASTM C231-04, Standard test method for air content of freshly mixed concrete by the pressure method.

RILEM Recommended Test Method: AAR-5—Detection of Potential Alkali-Reactivity—Rapid Preliminary Screening Test for Carbonate Aggregates

1 Foreword

The accelerated mortar-bar test (AAR-2) has been widely and successfully used as a screening test to identify aggregates that are potentially alkali-reactive. However there are some aggregates that, though alkali-reactive according to both field experience and concrete prism test results, do not show as such in the mortar-bar test. Trials carried out by RILEM have established that these aggregates include some carbonate materials and it has been found that such aggregates can be identified if 4/8 mm aggregate is used for making the bars instead of 0/4 mm aggregate. This modification of AAR-2 is designated AAR-5, and the international trials organised by RILEM to verify its effectiveness are described in [1]. The method is based on the Chinese Accelerated Concrete Microbar Method [2] developed by Nanjing University of Chemical Technology, China. However, use of the term 'concrete microbar' is not continued here, because this has caused confusion with the various concrete prism tests and also the very different French 'microbar' test.

The AAR-5 method is particularly beneficial when the aggregate contains both reactive silica and a kind of dolomite that is prone to dedolomitization. In the mortar-bar test (AAR-2) most of the dolomite (fine) aggregate will dedolomitize throughout, whilst in AAR-5 only the outer shells of the (coarser) aggregate is dedolomitized [3]. This difference seems to explain why, though alkali-silica gel develops in both the tests, the aggregate develops more pressure and expansion in AAR-5 than in AAR-2.

It is therefore recommended that, if dolomitic carbonate aggregates are being assessed, both the AAR-2 and the AAR-5 tests should be carried out. If the AAR-5 bars expand more than the AAR-2 bars, the reactivity of the aggregate is not of the normal alkali-silica type and further investigation using the 38 °C concrete prism test (AAR-3) or the 60 °C concrete prism test (AAR-4.1) will be required. The procedure for assessing potentially reactive carbonate aggregates is described in Annex A to AAR-0.

P.J. Nixon and I. Sims (eds.), *RILEM Recommendations for the Prevention of Damage by Alkali-Aggregate Reactions in New Concrete Structures*,
RILEM State-of-the-Art Reports 17, DOI 10.1007/978-94-017-7252-5_7

2 Scope

This test method is intended to be used conjunctively with AAR-2 (accelerated mortar-bar test method) to determine rapidly the potential alkali-reactivity of carbonate aggregates through the comparative evaluation of the expansion of mortar-bars, using 0/4 and 4/8 mm aggregates, immersed in NaOH solution at elevated temperature as specified in the methods.

Some carbonate aggregates do not show as potentially reactive in the AAR-2 test, though they are alkali-reactive both according to field experience and test results according to the AAR-3 (38 °C concrete prism method).

When used in conjunction with AAR-2, AAR-5 provides a means of reliably screening carbonate aggregates, in particular aggregates that contain dolomitic limestone and/or dolomite/dolostone, for their potential alkali-reactivity. It is recommended that the test is performed not only with the aggregate to be evaluated but also with a known reactive and a known non-reactive reference aggregate.

3 Principle

This method is similar to AAR-2: bars moulded from a mix containing the aggregate to be tested are demoulded after 24 h, heated up in water to 80 °C during another 24 h, then immersed in 1 M NaOH solution at 80 °C and the expansions are measured. However, since some carbonate aggregates produce deleterious expansions only if used in a larger particle size, 4/8 mm aggregate is used instead of 0/4 mm aggregate (see Figs. 1 and 2) and the 'short fat' 40/40/160 mm bars (Option AAR-2.2) are considered instead of the 'long thin' 25/25/285 mm bars (Option AAR-2.1).

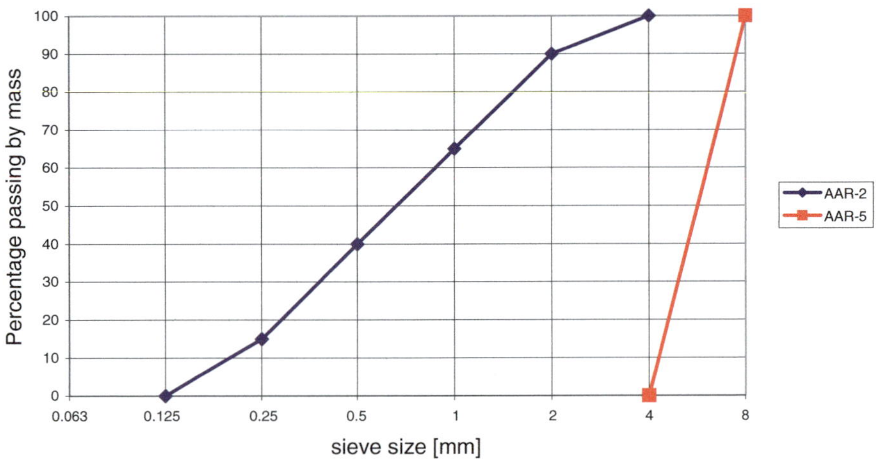

Fig. 1 Grading curves for AAR-2 & AAR-5 aggregates

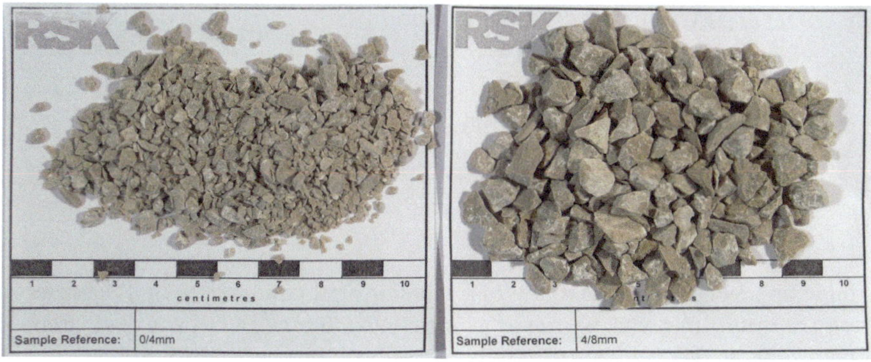

Fig. 2 Comparison of AAR-2 (*left*) and AAR-5 (*right*) aggregates

Note 1: *4/8 mm is the preferred aggregate size. If these sieves are not available and other size fractions are used (e.g. 5/10 mm as in the Chinese method* [2]) *expansions may be different, see also Annex* A3.

4 Apparatus

4.1 Crusher

A laboratory-type crusher of suitable size and design capable of crushing aggregate to the prescribed size fractions.

4.2 Sieves

A set of sieves conforming to series A of ISO 6274 [4], having square apertures of 8 and 4 mm.

4.3 Balance

A balance capable of weighing 1000 g with an accuracy of 1 g.

4.4 Measuring Cylinders

Graduated in ml, with a capacity of 500 ml.

4.5 Mixer, Paddle and Mixing Bowl

A mixer, paddle and mixing bowl, as used in the procedure for testing the strength of cement with plastic mortar (EN 196-1 [5] or ASTM C109 [6]).

4.6 Flow Table

A flow table consisting essentially of a circular rigid table top, with a diameter of about 300 mm, that can be raised vertically by means of a cam, and dropped through a nominal height of 10.0 mm (e.g. conforming to EN 1015-3 [7] or equivalent ASTM C230 [8]).

4.7 Mould for Flow Test

A frustum of a cone with height 60 mm, base diameter 100 mm and top diameter 70 mm (e.g. conforming to EN 1015-3 [7] or equivalent ASTM C230 [8]).

4.8 Tamper for the Flow Table

A tamper, made of non-absorbent, non-abrasive, non-brittle material and with a cross-section of 40 mm in diameter. A convenient length is 230–300 mm. The tamping face shall be flat and at right angles to the length of the tamper. The tamper mass shall be 0.250 kg (e.g. conforming to EN 1015-3 [7] or equivalent ASTM C1437 [9]).

4.9 Bar Moulds

Moulds, providing for bars with a nominal length of 160 mm and a cross section of 40 mm × 40 mm (Option AAR-2.2). The end plates of the moulds must have threaded holes in the centres to take stainless steel pins of 6 mm diameter and 20 mm length used for length measurements.

4.10 Tamper for the Bar Moulds

Tamper (not applicable if a vibrating table is used for compacting the specimens), made of non-absorbent, non-abrasive, non-brittle material and with a cross-section of 13 × 25 mm. A convenient length is 120–150 mm. The tamping face shall be flat and at right angles to the length of the tamper.

4.11 Vibrating Table

A vibrating table (not applicable if the specimens are compacted by hand) as used in the procedure for testing the strength of cement with plastic mortar.

4.12 Length Comparator

Consisting of:

- an apparatus to measure the length of the specimens conveniently and rapidly;
- a high-grade dial micrometer, or other measuring device, graduated to read in 1.0 μm units, accurate to within 2.0 μm in any 20 μm range, and within 5 μm in any 0.25 mm range. The measuring range shall allow for small variations (±10 mm) from the nominal gauge length of the specimens;
- an Invar® reference bar (or similar) of the same nominal length as the specimens for checking the measuring device, before and after each set of readings.

4.13 Containers

Rigid containers for the test specimens made of plastic or other material resistant to corrosion by a solution of sodium hydroxide at a temperature of 80 °C for a prolonged period of time. Each container must be of such dimension and shape to accommodate at least three specimens and must be provided with lids or other suitable means to prevent loss of moisture by leaking or evaporation. The bars must be positioned and supported in such a way that the solution has access to the whole of the bar. It should further be ensured that the specimens do not touch each other or the sides of the container. The specimens, if stood upright in the solution, must not be supported by the steel pins.

4.14 Storage

A cabinet or moist storage room maintained at a temperature of 20 ± 1 °C and a relative humidity more than 90 % (e.g. conforming to EN 196-1 [5] or equivalent ASTM C109 [6]).

An oven or room of suitable size to accommodate the required number of containers maintained at a temperature of 80 ± 2 °C.

5 Reagents and Materials

5.1 Water

Distilled or deionised water.

5.2 Sodium Hydroxide Solution (NaOH)

Each litre of sodium hydroxide solution shall contain 40.0 g of NaOH dissolved in 900 ml of water and, after cooling to about 20 °C, it shall be diluted with additional distilled or deionised water to obtain 1.0 l of solution. The concentration of the solution shall lie between 0.99 and 1.01 M. A new solution shall be prepared for each series of tests.

5.3 Cement

An ordinary Portland cement CEM I [10] or ASTM type I [11], with a minimum Na_2O equivalent (Na_2O + 0.658 K_2O) of 1.0 % shall be used. The specific surface of the cement, when measured according to the air permeability method (e.g. EN 196-6 [12] or ASTM C204 [13]), shall be greater than 450 m^2/kg. The autoclave expansion, determined according to ASTM test method C151 [14] shall be less than 0.20 %. Alternatively the MgO soundness can be evaluated through a Le Chatelier test (e.g. EN 196-3 [15]) and the increase in separation of indicator ends shall be 0 mm. Check the cement to see if any lumps are present. If these cannot be crushed between your fingers, sieve the cement on a 250 μm sieve.

Note 2: *The composition of the cement may influence the expansion, even when the alkali content of the mix is kept at the same level according to clause 7.3. Therefore, the same cement should be used both for the AAR-2 and the AAR-5 tests.*

Note 3: *A suitable reference cement is available from Norcem A.S (Norway): see RILEM Recommendation AAR-0.*

6 Procedure

6.1 Preparation of the Aggregate Sample

The quantity of the sample delivered to the laboratory should be in accordance with the sampling procedures recommended in AAR 1.1. The material to be tested shall consist of particles <8 mm and >4 mm, processed according to Sects. 6.1.1

and 6.1.2. A comparison between the aggregate gradings for AAR-2 and AAR-5 is shown in Figs. 1 and 2.

Note 4: *For the influence of the size of the aggregate, see Annex* A3.

6.1.1 Natural 4/8 mm Aggregate

The natural 4/8 mm aggregate shall be tested in the grading as submitted, except that the aggregate is washed, dried, and sieved and particles retained on an 8 mm screen and particles passing a 4 mm screen are removed.

6.1.2 Aggregate >8 mm

The aggregate >8 mm shall be crushed, sieved, washed, and dried to produce a 4/8 mm sample without oversize and undersize material as in Sect. 6.1.1.

6.1.3 Samples to be Tested

When a petrographic examination reveals that the natural 4/8 mm aggregate and the crushed 4/8 mm aggregate from the same source are essentially different, both the samples shall be tested; otherwise only the crushed aggregate needs to be tested.

6.2 Conditioning

The temperatures of the moulding room, apparatus, dry materials, mixing water and of the cabinet or moist storage room shall be maintained at 20 ± 1 °C. The relative humidity of the moulding room and of the cabinet or moist storage room shall not be less than 50 and 90 % respectively.

6.3 Proportioning of the Mix

Take 900 g of cement, 900 g of 4/8 mm aggregate, and 290 ml of water (see also Note 6) for each batch for three bars.

Before mixing, add enough sodium hydroxide to the water to bring the alkali content of the concrete up to a Na_2O equivalent of 1.5 % by mass of cement (see Annex A6.3).

If the mix is not sufficiently workable (e.g. when the surface of the aggregate is very rough), add just enough superplasticizer (but not one containing an air-entraining agent) as is necessary. In order to minimize any risk of segregation, flow values in excess of 150 mm (determined according to EN 1015-3) should be avoided. Any alkali content of the plasticizer (using manufacturer's declared data where available) shall be taken into account when calculating the Na_2O equivalent.

Note 5: *In the original Chinese method the water content was 270* ml. *In a comparative test conducted in the course of the international trial in 2002, a somewhat higher water content produced the same expansions. The water content of 290* ml *was chosen to improve the workability of the mix and to minimize the need for a superplasticizer. In the 2002 trial the superplasticizer did not influence the expansions, but it is thought that a high amount of a superplasticizer in combination with special cements may cause early stiffening.*

Note 6: *The water content of 290* ml *is the free water available for the workability of the fresh mix and for the hydration of the cement. The total water to be added to the mix is the free water plus the water absorbed by the aggregate to bring it to a saturated surface dry condition. The water absorption can be measured according to EN 1097-6* [16] *or equivalent ASTM C128* [17].

Note 7: *The water content of the superplasticizer should be included in the free water content. However, with amounts of superplasticizer not exceeding 10* g *there is no need to make allowance for its water content.*

Note 8: *In the original Chinese method KOH was added to the mix to bring up the alkali level to the required level. In comparative tests by RILEM the addition of NaOH resulted in somewhat higher expansions; NaOH was chosen for consistency with AAR-2.*

6.4 Moulding Test Specimens

Make at least three 40 × 40 × 160 mm specimens for each aggregate to be tested.

Prepare the specimen mould with a suitable releasing agent that will not affect the setting of the cement or leave any residue that will inhibit the penetration of water into the specimen.

Mix the constituents in accordance with the procedure prescribed in AAR-2 (see also Note 9). If necessary, superplasticizer has to be added at the end of the mix procedure, with further mixing.

The mould shall be filled in two approximately equal layers, each layer being compacted and levelled with the suitable tamper. Work the concrete along the surface of the mould with the tamper until a homogeneous specimen is obtained, making sure that the concrete is fully pushed under the reference inserts before a second layer is placed into mould. After the top layer has been compacted, cut off the concrete flush with the top of the mould and smooth the surface with a few strokes of the trowel.

Note 9: *The clearance between the mixer bowl and the paddle should be increased in order to avoid blocking and damage to the machine due to the coarse aggregate.*

Note 10: *If the bars are compacted by hand-tamping, normally no superplasticizer should be necessary. If a vibrating table is used for compacting the mix, the addition of a small amount (e.g. 5 g) of a superplasticizer may be necessary in order to enable proper compaction. A flow of at least 115 mm as determined according to EN 1015-3 has been found to ensure this.*

Note 11: *Particular care should be taken to attain a consistent compaction of the mix as the degree of compaction greatly influences the degree of expansion.*

6.5 Initial Curing and Measurement

Place the moulds in the moist cabinet or in the moist storage room (20 ± 1 °C and >90 % RH) for a period of 24 ± 2 h. Then, remove the specimens from the mould and, while they are being protected from loss of moisture, weigh and properly identify each specimen in such a way that they, when subsequently measured, are placed in the measuring equipment in the same manner. Make and record the initial length (Li) and all subsequent measurements to the nearest 0.001 mm.

Place the specimens made with each aggregate sample in a storage container with sufficient distilled or deionized water, at room temperature, to immerse them totally. Seal and place the containers in an oven at 80 ± 2 °C for a period of 24 h.

Remove the containers from the oven one at a time. Remove other containers only after the bars in the first container have been measured and returned to the oven. Remove the bars one at a time from the water and dry their surfaces with a towel or cloth paying particular attention to the two metal pins. Take the zero measurement of each bar (Lo) immediately after drying and read as soon as possible after the bar is in position. Complete the process of drying and measuring within 15 s of removing the specimens from the water. The measuring device should be checked with the reference bar prior to and after measurement of each set of specimens.

6.6 Final Storage and Measurement

Place the specimens made with each aggregate sample in a container with sufficient 1 M NaOH, preheated at 80 ± 2 °C, totally to immerse the specimens (1 example container is shown in Fig. 3). The recommended volume proportion of sodium hydroxide solution to concrete bars in a storage container shall be 4 ± 0.5 times the volume of the bars. Seal the container, mark the level of the liquid on the outside, and return it to the oven.

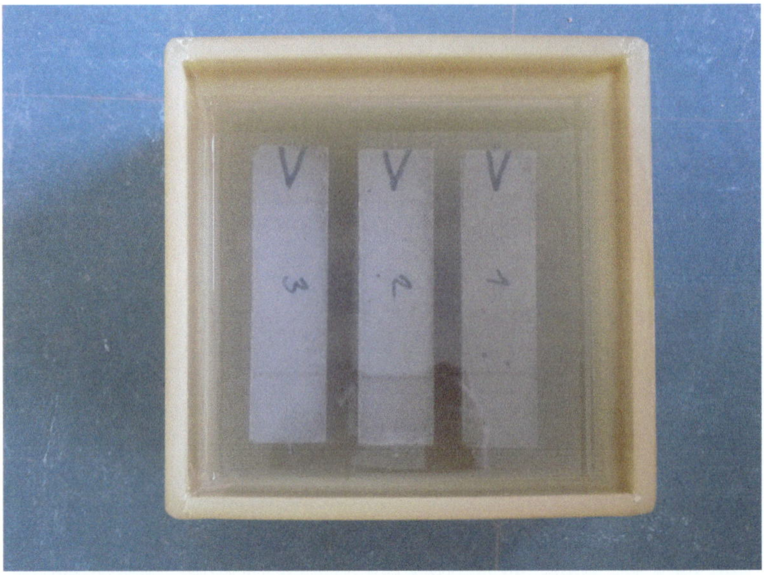

Fig. 3 Container with set of 40/40/160 mm bars submerged in NaOH-solution

Take subsequent measurements (Ln) of the specimens periodically, with a reading after 24 h of immersion in the NaOH solution and at least three intermediate readings before the final reading at 14 days (e.g. 3, 6 and 9). If so desired, measurements may be taken at 24-h intervals and may be continued beyond 14 days from the zero measurement (e.g. 21, 28 and 56 days). All measurements should be taken at approximately the same time each day. The measuring procedure is identical to that described in clause 6.5 and the specimens are returned to their container after each measurement. If moisture is lost from the container by evaporation replenish with distilled or deionized water.

Note 12: *Judgement will normally be made after storage for 14 days in NaOH. Continuation of storage and measurement until 28 or 56 days is recommended in order to collect data that may be useful for establishing criteria to predict AAR-3 results.*

6.7 Safety Measures

Reliable safety precautions should be taken and suitable personal protective equipment should always be used to avoid the hazards of the hot alkaline solution such as severe burns and injury to unprotected skin and eyes.

7 Calculation and Reporting of Results

7.1 *Expansion*

The gauge length, corresponding to the distance between the inner ends of the metals pins, has to be accurately evaluated after demoulding (when making the initial measurement Li), taking into account the length of the metal pins, these last being measured to the nearest of 0.1 mm.

The linear expansion of each specimen is obtained by calculating the difference between the length of the specimen at each period of measurement (Ln) and the zero measurement (Lo), to the nearest 0.001 % of the effective length, as follows:

$$\text{Expansion, \%} = 100 \times (Ln - Lo)/\text{Gauge length}$$

where:

Ln = reading taken at each period of storage in sodium hydroxide solution, n being the number of days counted from the zero measurement,
Lo = measurement of specimen just before immersion in sodium hydroxide solution (zero measurement),
Gauge length = distance between inner ends of the metal pins, measured to the nearest 0.1 mm.

Lo is taken after 1 day of storage of the bars in water bath at 80 ± 2.0 °C and just before their immersion in sodium hydroxide solution.

Ln is the measurement referred to the days of immersion in sodium hydroxide solution at 80 °C, starting immediately after the zero measurement. Then the measurement Ln = L_{14}, taken after 14 days of immersion, corresponds to the measurement taken 16 days after casting.

7.2 *Expansion Recording and Further Examination*

Record and report to the nearest 0.01 % the average expansion of the specimens for a given period. If a specimen breaks during the test, the test will be considered valid provided the two remaining specimens do not differ from each more than the values given below. For average values of expansion greater than 0.10 %, the repeatability is considered satisfactory if the expansion of each specimen is within 10 % of the average value, with a confidence level of 95 % (i.e. the probability of one case out of 20 being an incorrect result). For average values of expansion less than 0.10 %, the repeatability is considered satisfactory if, for each specimen, the deviation from the average value is within 0.01 %. If these values are exceeded, deem the test invalid and repeat the test.

After the final measurement, the specimens shall be examined and any relevant features recorded. Warping, if observed, shall be measured on 3 moulded surfaces by placing the specimens on a plane surface, with curved ends facing downwards, and measuring the maximum separation between the specimen and the surface to the nearest 0.2 mm. Location, type and pattern of any cracking should also be recorded.

Note 13: *If the bars exhibit an expansion that is judged to be deleterious, a petrographic examination of the interior of the bars that have expanded most should be carried out together with an examination of the internal crack pattern to confirm that the cause of expansion is likely to be an alkali-aggregate reaction. If it has been concluded from the expansion results and supplementary examination of the bars that a given aggregate should be considered potentially alkali-reactive, additional studies using RILEM Recommended test methods* [15], *may be appropriate to develop further information on its potential alkali-reactivity and to evaluate the effect of coarse aggregate, different aggregate gradings and different alkali contents of the concrete.*

8 Test Report

The following information shall be given in the report:

- identification and source of the aggregate sample and reference to petrographic analysis, if available; date of and condition at delivery at the laboratory;
- type and maximum size of the aggregate;
- type of processing undertaken on the aggregate sample in the laboratory (washing, drying, crushing, sieving etc.);
- grading of the aggregate as used in the test;
- identification and source of the Portland cement;
- alkali content of the cement expressed as equivalent sodium oxide (% Na_2O + 0.658 % K_2O);
- autoclave expansion and/or Le Chatelier value of the cement;
- Blaine fineness of the cement;
- workability of the mix (flow, if measured);
- type and content of superplasticizer (if used), and its contribution to the water and alkali contents of the mix;
- size and mass of the test bars;
- any tested reference (reactive and non-reactive) aggregates;
- initial expansion of the bars after 24 h of storage in water at 80 °C;
- each single value and average percentage length change after each measurement of the specimens;
- a graph of the percentage length change vs time from the zero reading to the end of the 14-day period of immersion in NaOH solution;
- results of any warping measurements of the specimens;
- any significant features revealed by examination of the specimens and the sodium hydroxide solution during and after the test.

Annex

A2 SCOPE—Reference Aggregates

Suitable reference aggregates, i.e. a reactive carbonate aggregate from Pittsburg quarry, Kingston, Ontario, Canada and a reference non-reactive aggregate, are described in AAR-0. As an alternative to the designated non-reactive reference aggregate, a natural gravel or crushed rock may be used that is known from experience to be non-reactive and, when tested according to AAR-2, exhibits an expansion of not more than 0.02 %.

A3 PRINCIPLE—Influence of the Size of the Aggregate

4/8 mm is the preferred aggregate size. If these sieves are not available and other size fractions are used (e.g. 5/10 mm as in the Chinese method [2]) expansions may be different, a larger upper size leading to a larger expansion if reactive carbonate material is present and a smaller upper size leading to smaller expansions. It is recommended that the upper aggregate size is not changed by more than 1 mm.

A6.3 Proportioning of the Concrete—Example of Calculation of NaOH Addition

In the original Chinese method a cement in which the alkali content was enhanced with NaOH to give the equivalent of 1.5 % Na_2O equivalent was used. As there is no experience at present of testing aggregates containing reactive carbonate material using lower alkali cements, that recommendation is presently retained in this method. It is hoped that with further experience this can be changed to allow the use of the same level of alkali in the cement as in AAR-2.

In order to bring the alkali content of the mix up to a Na_2O equivalent of 1.5 % by mass of cement, enough sodium hydroxide (NaOH) has to be added to the mixing water. An example of calculation is here reported, assuming that the ordinary Portland Cement available for the test be characterized by an alkali content of 1.0 %.

$$\begin{aligned}
\text{Cement content in the mix} &= 900\ \text{g} \\
Na_2O\ \text{eq. required in the mix (1.5 \% by mass of cement)} &= 900 \times 0.015 = 13.5\ \text{g} \\
Na_2O\ \text{eq. provided by the cement (1.0 \% by mass of cement)} &= 900 \times 0.010 = 9.0\ \text{g} \\
Na_2O\ \text{eq. to be added to the mixing water} &= 13.5 - 9.0 = 4.5\ \text{g}
\end{aligned}$$

where 1.291 is the conversion factor between Na_2O eq. and NaOH.

References

1. Sommer, H., Grattan-Bellew, P., Katayama, T., and Tang, M., Development and Interlaboratory Trial of the RILEM AAR-5 Rapid Preliminary Screening Test for Carbonate Aggregates, Tang, M, and Deng, M (eds), Proceedings 12th International Conference on Alkali-Aggregate Reaction in Concrete, Beijing, October 2004, Volume I, 407–412.
2. Xu, Z., Lan, X., Deng, M., and Tang, M., A new accelerated method for determining the potential alkali-carbonate reactivity, Proceedings of the 11th International Conference on Alkali-Aggregate Reaction, Quebec, 2000, pp 129 –138.
3. Katayama, T., How to Identify Carbonate Rock Reactions in Concrete, Proceedings of the 9th Euroseminar on Microscopy Applied to Building Materials, Trondheim, September 2003.
4. ISO 6274:1982, Concrete - Sieve analysis of aggregates.
5. EN 196-1:2005 (eg BS EN 196-1) Methods of testing cement - Determination of strength.
6. ASTM C109/C109 M-11a, Standard test method for compressive strength of hydraulic cement mortars (using 2-in. or [50-mm] cube specimens).
7. EN 1015-3:1999 (eg BS EN 1015-3), Methods of test for mortar for masonry – Determination of consistency of fresh mortar (by flow table).
8. ASTM C230/C230M-14: Standard specification for flow table for use in tests of hydraulic cement.
9. ASTM C1437-09, Standard test method for flow of hydraulic cement mortar.
10. EN 197-1:2011 (eg BS EN 197-1), Cement - Composition, specifications and conformity criteria for common cements.
11. ASTM C150/C150 M-11, Standard specification for Portland cement.
12. EN 196-6:2010 (eg BS EN 196-6) Methods of testing cement: Determination of fineness.
13. ASTM C204-11: Standard test methods for fineness of hydraulic cement by air-permeability apparatus.
14. ASTM C151/C151 M-09, Standard test method for autoclave expansion of hydraulic cement.
15. EN 196-3:2005 + A1:2008 (eg BS EN 196-3), Methods of testing cement - Determination of setting time and soundness.
16. EN 1097-6:2013 (eg BS EN 1097-6): Tests for mechanical and physical properties of aggregates: Determination of particle density and water absorption.
17. ASTM C128-07a, Standard test method for density, relative density (specific gravity), and absorption of fine aggregate.

RILEM Recommended Specification: AAR-7.1—International Specification to Minimise Damage from Alkali Reactions in Concrete—Part 1: Alkali-Silica Reaction

1 General

This Part 1 of AAR-7 principally addresses aggregates containing reactive silica.

Part 2 of AAR-7 provides separate guidance on reactive carbonate aggregates.

Part 3 of AAR-7, which is currently available in preliminary form, provides specific guidance on very large, long-service structures, such as dams.

Alkali-Silica Reaction (ASR) in concrete is a result of reaction between the alkaline pore solution in concrete and reactive silica in the aggregate. The reaction leads to the formation of a gel which can absorb water and exert an expansive force on the concrete. In certain conditions these reactions can lead to damaging expansions and cracking in the concrete. For such damaging expansion to occur all of the following conditions must be present simultaneously:

- A sufficiently alkaline pore solution
- A critical amount of reactive silica
- A sufficient supply of water

Effective specifications to avoid damage from the reaction are based on ensuring that at least one of these conditions is absent.

Additionally, there are interactions with other environmental actions such as freezing and thawing, application of de-icing salts and exposure to a marine atmosphere for which allowance must be made.

Such specifications can result in greater costs and in adverse environmental effects, for example by restricting the choice of aggregates or disposal of alkaline cement kiln dust, so it is also important to tailor the precautions to the nature and service life of the structure.

Therefore, the development of the precautions should take the following form:

1. Determination of the necessary level of precaution;
2. Undertaking recommendations according to the level of precaution required.

P.J. Nixon and I. Sims (eds.), *RILEM Recommendations for the Prevention of Damage by Alkali-Aggregate Reactions in New Concrete Structures*, RILEM State-of-the-Art Reports 17, DOI 10.1007/978-94-017-7252-5_8

The assessment of the level of precaution and the appropriate precautionary measures are summarised in Fig. 1.

Note 1: *This flow chart may not apply without modification to dams and other similar structures (see Sect.* 3.4*).*

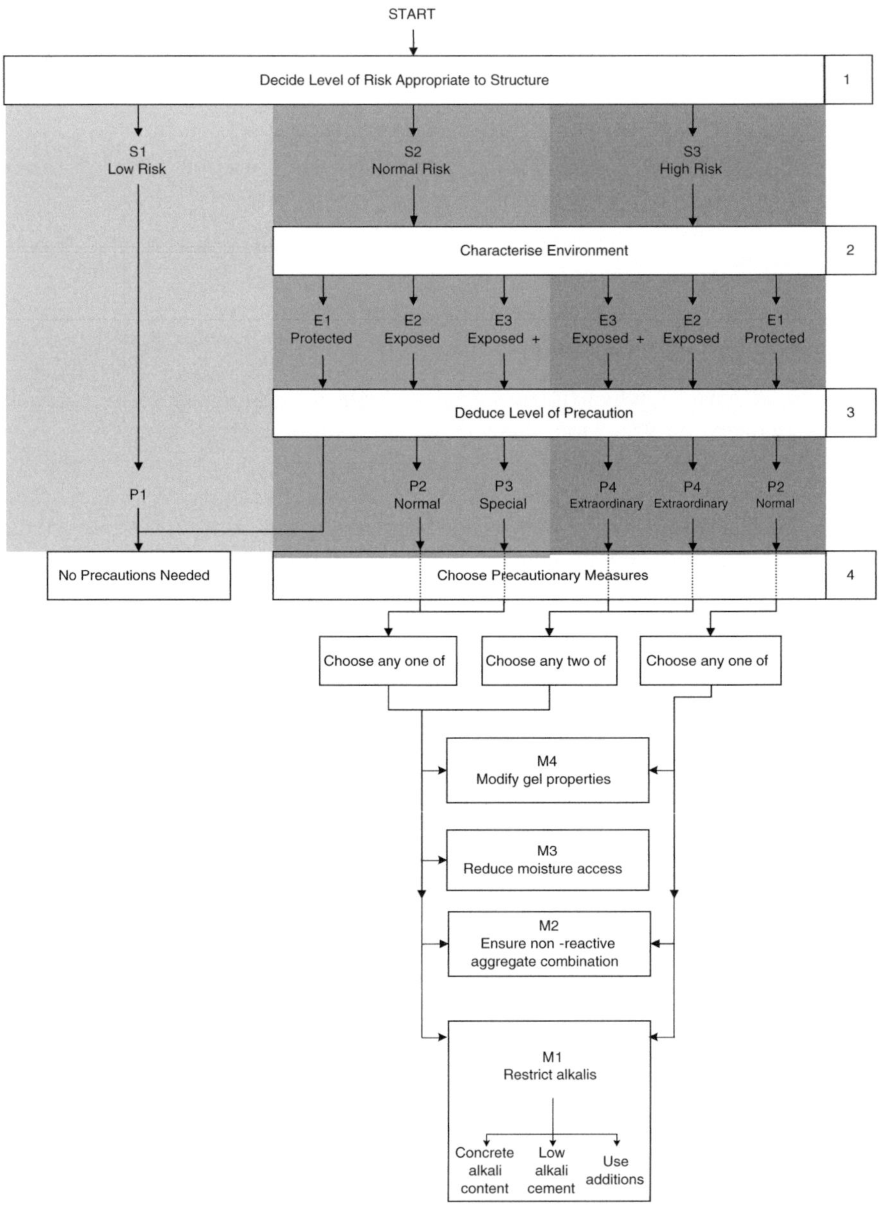

Fig. 1 Flow chart summarising the specification of siliceous aggregates

2 Determining the Level of Precaution

The key activities here are the characterisation of the structural needs and service life and the characterisation of the environment.

2.1 Structural Needs and Service Life

Damaging alkali-aggregate reactions are slow and progressive. Typically in temperate and cooler climates they begin to cause visible damage after 5–10 years and then may continue for 20 or 30 years or longer; in warmer climates the reactions are accelerated. Some slowly reacting aggregates, however, result in deterioration that takes much longer but can eventually be more destructive. The damage is evidenced as cracking and expansion of the concrete, occasionally leading to displacement of individual elements in a structure. While this has structural consequences, particularly if the reinforcement is insufficient, any deterioration will normally be evident well before there is danger to the integrity of the structure.

Three levels of categorisation of structure, according to the risks associated with any deterioration and the consequent need for precautions, are therefore appropriate:

S1—low risk
S2—normal risk
S3—high risk

It is the responsibility of the owner, or authority responsible for the structure, to decide on the appropriate level of risk in co-operation with the designer. This decision will be affected by the economic effects of any failure or deterioration as well as engineering and safety considerations. Other factors to be taken into account are the ease with which any deterioration can be detected, monitored and managed, the importance of the appearance of the structure and likely public perceptions of safety. Criteria which will assist in making this decision are given in Table 1.

2.2 Characterisation of the Environment

When all the necessary compositional factors are present, the likelihood and extent of damaging alkali-silica reaction is dependent above all on the supply of moisture. In the majority of cases, a supply of moisture extraneous to the concrete itself is necessary. Other, aggravating, factors which will influence the likelihood of damage and its severity include the application of sodium chloride based de-icing salts, exposure to seawater and the synergistic effects of freezing and thawing damage. In concrete roads the stress variation caused by fluctuating loads may also be an aggravating factor.

Table 1 Structures classified by risk category

Category-consequences of damage	Acceptability of ASR damage	Examples
S1 Safety, economic or environmental consequences of deterioration small or negligible	Some deterioration from ASR is acceptable	• Non load-bearing elements inside buildings • Temporary or short service life structures (likely design life 10–20 years) • Small numbers of easily replaceable elements • Most low-rise domestic structures
S2 Some safety, economic or environmental consequences if major deterioration	Minor ASR damage is acceptable/manageable	• Most building and civil engineering structures • Precast elements where economic costs of replacement are severe; e.g. railway sleepers • Normally designed for service life up to 100 years
S3 Serious safety, economic or environmental consequences if any deterioration	No significant damage acceptable	• Long service life (+100 years) or highly critical structures/elements where the risk of deterioration from AAR damage is judged unacceptable, such as • Nuclear installations, dams, tunnels • Exceptionally important bridges or viaducts • Structures retaining hazardous materials • Exceptionally critical elements impossible/very difficult to inspect or replace/repair • Structures where the economic risk of non-serviceability would be unacceptable

Three levels of categorisation of environment are therefore appropriate:

E1. The concrete is essentially protected from extraneous moisture

E2. The concrete is exposed to extraneous moisture

E3. The concrete is exposed to extraneous moisture and additionally to aggravating factors, such as sodium chloride based de-icing salts, freezing and thawing or wetting and drying in a marine environment

More detail on the factors affecting the environmental categorisation is given in Table 2.

Table 2 Environmental classes

Environmental class	Description	Environment of concrete (see table Notes)
E1	Dry environment protected from extraneous moisture	• Internal concrete within buildings in dry (1) service conditions
E2	Exposed to extraneous moisture	• Internal concrete in buildings where humidity is high; e.g. laundries, tanks, swimming pools • Concrete exposed to moisture from the external atmosphere, to non-aggressive ground or immersed in water. • Internal mass concrete should be included in this category (2)
E3	Exposed to extraneous moisture plus aggravating factors	• Internal or external concrete exposed to deicing salts • Concrete exposed to wetting and drying by seawater (3) or to salt spray • Concrete exposed to freezing and thawing whilst wet • Concrete subjected to prolonged elevated temperatures whilst wet • Concrete roads subject to fluctuating loads

Notes
1. A dry environment corresponds to an ambient average relative humidity condition lower than 75 % (normally only found inside buildings) and no exposure to external moisture sources.
2. A risk of alkali-silica reaction exists for mass concrete elements in a dry environment because the internal concrete may still have a high relative humidity. Vulnerable mass concrete elements are those with a least dimension of 1 m or more.
3. Concrete constantly immersed in seawater does not suffer a higher risk of ASR than a similar element exposed to humid air, buried in the ground, or immersed in pure water, because the alkali concentration of sea water is lower than the alkali concentration of the pore solution of most concretes, and the penetration of chloride ions is usually limited to a few centimetres.

2.3 *The Level of Precaution*

The structural and environmental categorisation is combined into the level of precaution in Table 3, where four levels of precaution are identified:
P1. No special precautions against AAR
P2. Normal level of precaution
P3. Special level of precaution
P4. Extraordinary level of precaution

Table 3 Determination of level of precaution

	Environment category (see Table 2)		
	E1	E2	E3
Category of structure (see Table 1)	Level of precaution		
S1	P1	P1	P1
S2	P1	P2	P3
S3	P2	P4	P4

3 Specifications Appropriate to the Level of Precaution

The following precautionary measures may be applied:
(see Sect. 4 for more details).

M1: Measures to restrict the alkalinity of the pore solution
M2: Measures to ensure the use of a non-reactive aggregate combination
M3: Measures to reduce the access of moisture and maintain the concrete in a sufficiently dry state
M4: Measures to modify the properties of any gel such that it is non-expansive

Note 2: *Precautionary measures M1 and M2 are well established and are recommended by RILEM.*

Note 3: *The efficacy of M3 is difficult to assure, particularly in wetter climates, and it is recommended that it should not be applied as the only precautionary measure in S3 structures (see note in Sect.* 4.3).

Note 4: *M4 is newly developed and should be applied with caution.*

Note 5: *A reliable performance test for the concrete mix to be used could incorporate M1, M2 and M4. RILEM is developing such a test.*

3.1 Level of Precaution P1

No special precautions against AAR damage are necessary.

Ensure that appropriate standards and guidance are followed for the specification of the concrete and good practice employed in its placing and curing.

Note 6: *If this level of precaution is adopted, some damage from ASR is possible. Therefore the structure must be able to withstand this and the level of damage must be acceptable to the owner.*

3.2 Level of Precaution P2

This normal level of precaution against AAR damage is appropriate to structures where minor ASR damage is acceptable or damage can be monitored and managed. In precaution level P2 one of the precautionary measures M1, M2, M3 or M4 should be applied in the case of S2 (normal risk) structures, or one of the precautionary measures M1, M2 or M4 in the case of S3 (high risk) structures.

3.3 Level of Precaution P3

This special level of precaution is appropriate where minor ASR damage is acceptable or damage can be monitored and managed but where the structure is exposed to aggravating factors such as de-icing salts, freezing and thawing or wetting and drying in a marine environment. In such cases one of the precautionary measures M1–M4 should be applied and additionally the concrete should be designed to resist the aggravating factor; e.g. it should be freeze/thaw resistant or it should resist the ingress of de-icing salts or seawater.

3.4 Level of Precaution P4

This extraordinary level of precaution is only needed in structures where the consequences of any deterioration are unacceptable. In general it will necessitate the combined application of at least two of the precautionary measures M1–M4 and additionally the concrete in environmental class E3 should be designed to resist any aggravating factors such as freezing and thawing whilst wet, de-icing salts or wetting and drying in a marine atmosphere. This level should not be specified without careful thought and good reason, as it will almost certainly result in increased construction costs.

In some large/remote structures such as dams, where for environmental and/or economic reasons, it is necessary to use local materials, it may not be possible to apply two separate precautionary measures. In that case extra protection can be obtained by the more rigorous application of one of the precautionary measures; e.g. the use of a low alkali limit (see Sect. 4.1.1) and the inclusion of a fly ash or slag (see Sect. 4.1.2), or the use of a non-reactive aggregate combination (see Sect. 4.2) subject to a lower acceptance limit in the testing. Additionally, extra protection can be achieved through design and construction measures, such as detailing of the reinforcement, weather protection of critical elements, drainage or inclusion of expansion joints. More specific guidance for such structures is given in AAR-7.3, which is now available in preliminary form.

4 The Precautionary Measures

4.1 Precautionary Measure M1: Limiting the Alkalinity of the Pore Solution

There are various ways of achieving this:

- Limiting the alkali content of the concrete;
- Use of a low alkali cement;
- Inclusion of a sufficient proportion of a low lime-fly ash, other pozzolana demonstrated to be effective, or ground granulated blastfurnace slag in the concrete.

4.1.1 Limitation of the Alkali Content in the Concrete

In many respects this is the most easily applied and monitored of the measures. As it is now well demonstrated that the alkalinity of the pore solution is primarily dependent on the alkali content of the concrete mix, it is also the method recommended by RILEM to achieve Precautionary Measure M1.

Alkalis in Portland cement

The principal source of alkalis that control the alkalinity of the concrete pore solution is the alkali content of the Portland cement. This is usually expressed as the equivalent sodium oxide content:

$\%\ Na_2O$ equivalent = $\%\ Na_2O + 0.658\ \%\ K_2O$

Alkalis in concrete

The alkali content of the concrete is then the equivalent alkali content of the cement multiplied by the cement content of the mix plus any other reactive alkalis that should be included:

Concrete alkali (kg/m^3) = cement alkalis (%) × cement content (kg/m^3) + other sources of alkali

The effective application of this measure requires the support of the cement manufacturing and/or supply industry in a particular country, as it needs both the quality assured declaration of the average cement alkali level from a particular works and a measure of its variability. This is now done in several countries (e.g. France, UK) as part of the control and standardisation process of concrete. Alternatively users will need to assess cement alkali content by the reliable independent analysis of representative samples of the cement.

Other sources of alkali

There is no universal consensus on what other sources should be included. The general practice is to include alkalis from these sources:

- Residual salt in marine aggregates;
- Cementitious additions such as fly ash and slag;
- Admixtures;
- Mix water, especially if not from potable sources.

The proportion of alkalis that should be included from fly ash and slag is dependent on the proportion used and their reactivity (see Sect. 4.1.2). In a blended cement containing fly ash or slag, the manufacturer will need to declare an "effective" alkali content, calculated from the alkali in the Portland cement and the alkalis to be included (if any) from the fly ash or slag on the basis of Sect. 4.1.2.

Note 7: *Such manufacturers declarations are already available in some countries.*

Admixtures, particularly some superplasticisers, can contain substantial alkalis. The calculation of the alkali in the concrete will need to be based on the effective alkali declared by the manufacturer.

More problematic is the question of alkalis originating in the body of the aggregate. Many aggregates contain alkalis, but the extent to which these are released and contribute to the pore solution alkalinity remains uncertain and possibly variable. A method for assessing the potentially releasable alkali content of aggregates is being developed by RILEM as AAR-8.

At present, it is recommended that, unless there is national evidence to the contrary, this source of alkalis should not be included in the calculation except in concrete in S3 structures. In such structures the possibility of alkali release by the aggregate should be considered when using aggregates containing significant amounts of altered and/or weathered feldspar, or other minerals capable of releasing alkalis.

There is good evidence that the use of sodium chloride based de-icing salts can exacerbate ASR reactions. However, it is not possible to quantify this effect in order to include it in the calculation of the alkalis in the concrete. Instead, it is taken into account in deciding the environmental class of the concrete where it is one of the aggravating factors that lead to classification of the environment as E3.

Note 8: *There is recent evidence from the USA that de-icing salts containing sodium or potassium formates and acetates can cause ASR-like damage to concrete, even that containing 'non-reactive' aggregates.*

Alkali limits and aggregate reactivity

These must be set by national guidance. To enable the most efficient use to be made of the aggregate resources available in a particular country, the RILEM methods should be used to identify levels of aggregate reactivity that can then be used to set alkali limits (AAR-0). Aggregates that are essentially non-reactive can be identified by petrographic methods (AAR-1). The reactivity of more reactive materials is best defined in terms of the alkali threshold at which damaging ASR is first identified. To establish this threshold, the AAR-3.2 concrete prism method can be used. Alternatively, a particular country can designate the reactivity of particular rock types on the basis of their geology and known behaviour on concrete structures.

Three levels of aggregate reactivity can usefully be differentiated:

1. Low reactivity aggregates:
   These correspond to the aggregates identified as "non-reactive" in AAR-0;
2. Medium reactivity aggregates:
   These are aggregates that fall neither into the low nor high reactivity categories. They may well be the majority of aggregates in some countries, e.g. the UK (where these medium reactivity aggregates are termed "normal reactivity"), and include the siliceous sands and gravels that are found widely. They will have alkali thresholds of exceeding 4.0 kg/m^3 sodium oxide equivalent (Na_2O eq.);
3. High reactivity aggregates:
   This category corresponds to the aggregates shown by tests using concrete specimens (AAR-3) to have low alkali thresholds; typically 4.0. kg/m^3 Na_2O eq. or less. Such aggregates are likely to be rare.

Note 9: *Aggregates containing substantial proportions of opal are likely to be even more reactive than high reactivity aggregates and should either not be used in concrete or only with special precautions that have been proven effective by trials or performance tests* etc. *Such aggregates fall outside these recommendations.*

Note 10: *When classifying the reactivity of an aggregate it is important to assess the whole aggregate combination in order to allow for any "pessimum" effects (see Sect.* 4.2).

Examples of using the aggregate reactivity categories to set alkali limits are shown in Table 4.

Note 11: *The alkali limits in the example given in Table* 4 *allow a "safety margin" compared with the alkali thresholds determined in the laboratory (e.g. if the determined threshold was 4.0 kg/m^3, in this example the limit has been set at 3.0 kg/m^3: i.e. a 1.0 kg/m^3 safety margin). This is to allow for the known differences between laboratory and field specimens, experimental uncertainty (e.g. from alkali leaching) and site batching variability. National specifications making use of such limits will need to decide on an appropriate "safety margin" from local experience.*

Where, in a particular country, the alkali contents of the cements used are within a known narrow range, restrictions on the cement content according to aggregate reactivity and environment, can be used as an alternative method of limiting the alkali levels in the concrete mix. This can be easier for the concrete producer to apply, but care must be taken as the compositions of the cements may change over time because of, for example, changes in raw materials or import of cements.

Table 4 Alkali limits and aggregate reactivity

Aggregate reactivity	Alkali limit (kg/m^3 Na_2O eq.)
Low	None required
Medium	Typically 3.0–3.5 kg/m^3
High	Typically 2.5–3.0 kg/m^3

Low Alkali Cements

Although less precisely connected with the alkali concentration in the concrete pore solution, this is the longest established and most pragmatic countermeasure. In the case of low alkali Portland cements an upper limit of 0.60 % Na_2O equivalent is generally applied. The use of such low alkali cements has been found to be effective in some regions in preventing AAR damage, although there are occasional reports of damage despite their use and there is evidence from field trials in North America that with some aggregates damaging expansion can occur in concrete specimens made with low alkali cements.

Drawbacks are that cements guaranteed to meet this limit will often be more costly and, to achieve it, the manufacturer may have to discard kiln dust with adverse environmental effects. Moreover, this measure may not be effective in the case of concretes with unusually high cement contents, if there are significant sources of internal or extraneous alkali, or if the passage of moisture concentrates the alkalis in certain parts of the structure.

Some countries have also designated low alkali slag cements. In Germany, for example, two types of low alkali blastfurnace slag cement are defined, the allowable alkali content depending on the percentage of slag. In Austria there is good experience of using Portland-slag cements, compliant with particular Austrian standards, which have quite high alkali levels. With local experience, such approaches can be an effective way of using these supplementary cementing materials in avoiding ASR damage.

4.1.2 Use of Fly Ash, Slag, Other Pozzolanas and Other Mineral Additions

The use of concrete containing such additions, and of cements in which they are interground or mixed during manufacture, has been the subject of much controversy and research. It is clear that some of these materials can be very effective in combating AAR damage. However, their variability, internationally, makes it problematic to give specific RILEM Recommendations. The following advice summarises the general consensus ascertained by RILEM.

Low lime fly ashes, e.g. to EN 450 [1] or ASTM C618 class F [2], and ground granulated slags that are well established as effective cementitious materials or

Table 5 Recommended minimum fly ash and slag proportions

Aggregate reactivity	Low lime fly ash (<8 % CaO and <5 % Na_2O eq.)	Medium lime fly ash (8–20 % CaO and <5 % Na_2O eq.)	Ground granulated blastfurnace slag (<1.5 % Na_2O eq.)
	% by mass of total cementitious material		
Low	Any	Any	Any
Medium	25	30	40
High	40	Not recommended	50

constituent materials of the cement, will provide effective protection against AAR damage provided a sufficient proportion (as a proportion of the total cementitious material) is used. The proportion necessary will depend on the reactivity of the aggregate (Table 5).

Provided that these minimum proportions are used, and subject to local experience with particular materials, the alkali content of the fly ash or slag need not be included in the calculation of the "reactive" alkalis in the concrete.

If lower proportions of addition are used, some countries (e.g. UK, France) recommend that a proportion of the alkali content of the ash or slag is included in the calculation of the alkali content of the concrete mix. Based on these, a tentative RILEM recommendation is given in Table 6.

There is also good evidence that other highly active pozzolanic additions such as silica fume and metakaolin can be effective in protecting against ASR damage. It is tentatively recommended that for concrete containing medium reactivity aggregates the following minimum proportions are used:

Silica fume (>85 % SiO_2): 8 %
Metakaolin (>45 % SiO_2): 15 %

At present this approach is not recommended for concrete containing high reactivity aggregates.

Note 12: *It is vital that such materials are well dispersed in the concrete. Agglomerations of silica fume have caused damaging ASR expansions.*

Table 6 Alkali contributions from lower proportions of additions than are recommended in Table 5

Proportion of addition in cement	Proportion of alkali from addition to include in calculation of alkali content of concrete mix
Slag	
25–39 %	50 %
<25	100 %
Fly ash	
20–24 %	20 %
<20	100 %

Otherwise, sources of fly ash (e.g. high lime {>20 % CaO}) or slag that are not well established, or other pozzolanic materials or mineral additions, should only be used if their performance has been established by a concrete performance test based on, for example, the AAR-3 method. It is preferable that such performance tests are backed up with long-term outdoor field tests or by site experience, since some accelerated laboratory tests, using elevated humidities and temperatures, may give unrealistically optimistic results for the effectiveness of some pozzolanic materials in combating ASR expansion.

Note 13: *Such a performance test is being developed by RILEM.*

4.2 Precautionary Measure M2: Ensuring the Use of a Non-reactive Aggregate Combination

In the context of ASR, reactive silica occurs almost exclusively in the aggregate. Therefore to make use of this precautionary measure, the RILEM Recommended methods should be used, in accordance with AAR-0, to identify "non-reactive aggregate" combinations.

The proportion of silica that can lead to the most damaging reaction will depend on the reactivity of the silica. A small amount of highly reactive silica in the aggregate will be most damaging, whereas, if the aggregate contains a high proportion of such highly reactive silica, there may be little damage. If an aggregate containing highly reactive silica is mixed with a non-reactive one, the behaviour of the mix will vary from very damaging to not damaging at all, depending on the proportions of the mix. This feature is known as the "pessimum" effect. Conversely, in aggregates containing low reactivity forms of silica or where the silica is not easily exposed to the alkaline pore solution, the worst damage may occur when the greatest amount of silica is present.

Because of this, it is important that the whole aggregate combination is assessed, as amounts of reactive silica that are innocuous in either the fine or coarse aggregate alone may be damaging in the combined aggregate. Conversely, apparently reactive fine or coarse aggregates may be safe when used in combination. Both AAR-3 and AAR-4 are suitable for assessing the combined aggregate. When used correctly, controlling this 'pessimum' effect can be an effective way of combating AAR damage and making use of available natural resources.

4.3 Precautionary Measure M3: Reducing Moisture Access and Maintaining the Concrete Sufficiently Dry

This can be achieved at the design stage by the use of, for example, external cladding or tanking (protection using a completely waterproof barrier). Inclusion of well designed drainage that can be inspected and maintained is also important. Use of cladding on a concrete structure in a cold and very humid environment may however not sufficiently reduce the moisture content in the air behind the cladding. The use of cladding is regarded as most feasible in a warm and dry environment.

The effectiveness of surface treatments (coatings and impregnations) in preventing (or in arresting) AAR damage has been found to be variable, and thus surface treatment is not regarded as a sufficient precautionary measure.

Note 14: *If a surface treatment is applied as an extra precaution measure, it should be vapour permeable so that the concrete can dry out when the humidity is low. The maintenance of any coating and/or drainage measures used is vital.*

4.4 Precautionary Measure M4: Modifying the Properties of the Gel Such that It Is Non-expansive

Inclusion of sufficient, soluble, lithium salts in the concrete mix water is an effective means of counteracting AAR damage. It is believed to operate by modifying the nature of the ASR gel such that it does not absorb water and exert an expansive force.

Lithium nitrate is the preferred salt. The recommended dosage levels depend on the alkali level in the concrete and the nature of the aggregate. North American research [3] suggests that lithium salts are not equally effective with all reactive aggregate types. Some aggregates require much higher doses of lithium than others, whilst expansion with other reactive aggregates can sometimes be controlled with lower lithium doses than have previously been recommended. Unlike other measures based on using additions, such as fly ash or ground granulated blastfurnace slag, the lithium dose required does not appear to be related to the degree of aggregate reactivity. Accordingly, it is not possible at present to recommend a single dosage of lithium nitrate that will be effective for all aggregates. Instead, performance testing to evaluate its effectiveness and determine an appropriate lithium dosage is recommended for those considering the use of precautionary measure M4 in new concrete.

Note 15: *A suitable performance test is being developed by RILEM. Until that is available, North American experience suggests evaluation using a version of AAR-3 with an acceptance criterion of 0.04 % after two years. Also, a suitable performance test based on a modified version of the AAR-2 type of accelerated mortar-bar test is currently under development in North America.*

References

1. EN 450-1:2012 & EN 450-2:2005, Fly ash for concrete, Part 1: Definition, specifications and conformity criteria (+Amendment A1:2007), & Part 2: Conformity evaluation.
2. ASTM C618-08A, Standard specification for coal fly ash and raw or calcined natural pozzolan for use in concrete, *American Society for Testing and Materials*, West Conshohocken, Pennsylvania, USA.
3. Tremblay, C, Berube, M-A, Fournier, B, Thomas, M D A, Folliard, K J, 2007, Effectiveness of lithium-based products in concrete made with Canadian natural aggregates susceptible to alkali-silica reactivity, *ACI Materials Journal*, Title 104-M23, 104 (2), 195–205.

RILEM Recommended Specification: AAR-7.2—International Specification to Minimise Damage from Alkali Reactions in Concrete—Part 2: Alkali-Carbonate Reaction

1 General

Carbonate aggregates are generally not alkali-reactive. However, there are some circumstances in which carbonate aggregates appear to exhibit alkali-reactivity. These are:

- Carbonate rocks containing finely divided inclusions of silica. These may be termed siliceous limestones and can be very reactive towards alkalis. Because of the finely disseminated nature of the silica it can be difficult to detect by conventional optical petrography. The specification for concrete containing such rocks is dealt with in AAR-7.1. The expansive reactions are alkali-silica reactions and the measures discussed in AAR 7.1 are appropriate to control them;
- Some dolomitic carbonate rocks have been found to be expansive in the presence of alkalis, although pure dolostones[1] are not susceptible in this way. The mechanism of expansion is still a matter of controversy. One mechanism that has been advanced to explain the expansion is that the alkaline environment causes the breakdown (dedolomitization) of the dolomite crystals with accompanying precipitation of brucite and calcite [1]. However, research into some carbonate rocks has suggested that the deleterious expansion of dolomitic aggregates is caused by an alkali-silica reaction of cryptocrystalline quartz which is not detectable by optical microscopy [2]. According to this mechanism, dedolomitization is associated with the reaction but does not cause the expansion.

Regardless of the mechanism, the options for controlling alkali reactions in such dolomitic carbonate aggregates are much more limited than is the case for alkali-silica reactions. The primary strategy, given the present level of knowledge

[1]Dolostone: term synonymous with the traditional *Dolomite* rock—a rock composed entirely of the mineral dolomite ('dolostone' avoids confusion between dolomite rock and the mineral dolomite) [*Chambers Dictionary of Science and Technology*, 1999, Edinburgh, 1325 pp].

P.J. Nixon and I. Sims (eds.), *RILEM Recommendations for the Prevention of Damage by Alkali-Aggregate Reactions in New Concrete Structures*, RILEM State-of-the-Art Reports 17, DOI 10.1007/978-94-017-7252-5_9

and understanding is focussed on identifying and avoiding the use of such aggregates. However, experience suggests that some dolomitic carbonate aggregates may be useable in concrete in which the cement contains high replacement levels of blastfurnace slag.

Recommendations for the specification of concrete to avoid damaging reactions when using such dolomitic aggregates are given here in AAR 7.2.

Overall, therefore, the development of the precautions against any form of alkali reaction should take the following form:

1. Determination of the necessary level of precaution;
2. Undertaking recommendations according to the level of precaution required.

2 Determining the Level of Precaution

The key activities are the characterisation of the structural needs and service life and the characterisation of the environment.

2.1 Structural Needs and Service Life

Damaging alkali-aggregate reactions are slow and progressive. Typically in temperate and cooler climates they begin to cause visible damage after 5–10 years and then may continue for 20 or 30 years or longer; in warmer climates the reactions are often accelerated. Some slowly reacting aggregates, however, result in deterioration that takes much longer but can eventually be more destructive. In the case of reactions involving potentially susceptible carbonate aggregates, even in concrete containing a low alkali cement, expansion and damage to structures has become apparent in as little as 3 years and there has been damage to structures at 5 years. The damage is evidenced as cracking and expansion of the concrete, occasionally leading to displacement of individual elements in a structure. While this has structural consequences, particularly if the reinforcement is insufficient, any deterioration will normally be evident well before there is danger to the integrity of the structure.

Three levels of categorisation of structure, according to the risks associated with any deterioration and the consequent need for precautions, are therefore appropriate:

S1—low risk
S2—normal risk
S3—high risk

It is the responsibility of the owner, or authority responsible for the structure, to decide on the appropriate level of risk in co-operation with the designer. This decision will be affected by the economic effects of any failure or deterioration as well as engineering and safety considerations. Other factors to be taken into account are the ease with which any deterioration can be detected, monitored and managed, the importance of the appearance of the structure and likely public perceptions of safety. Criteria which will assist in making this decision are given in Table 1.

Table 1 Structures classified by risk category

Category—consequences of damage	Acceptability of ASR damage	Examples
S1 Safety, economic or environmental consequences of deterioration small or negligible	Some deterioration from ASR is acceptable	• Non load-bearing elements inside buildings • Temporary or short service life structures (likely design life 10–20 years) • Small numbers of easily replaceable elements • Most low-rise domestic structures
S2 Some safety, economic or environmental consequences if major deterioration	Minor ASR damage is acceptable/manageable	• Most building and civil engineering structures • Precast elements where economic costs of replacement are severe; e.g. railway sleepers • Normally designed for service life up to 100 years
S3 Serious safety, economic or environmental consequences if any deterioration	No significant damage acceptable	Long service life (+100 years) or highly critical structures/elements where the risk of deterioration from AAR damage is judged unacceptable, such as: • Nuclear installations, dams, tunnels • Exceptionally important bridges or viaducts • Structures retaining hazardous materials • Exceptionally critical elements impossible/very difficult to inspect or replace/repair • Structures where the economic risk of non-serviceability would be unacceptable

2.2 Characterisation of the Environment

When all the necessary compositional factors are present, the likelihood and extent of damaging alkali-aggregate reaction is dependent above all on the supply of moisture. In the majority of cases, a supply of moisture extraneous to the concrete itself is necessary. Other, aggravating, factors which will influence the likelihood of damage and its severity include the application of sodium chloride based de-icing salts, exposure to seawater and the synergistic effects of freezing and thawing damage. In concrete roads the stress variation caused by fluctuating loads may also be an aggravating factor.

Three levels of categorisation of environment are therefore appropriate for alkali aggregate reactions in general:

E1. The concrete is essentially protected from extraneous moisture
E2. The concrete is exposed to extraneous moisture
E3. The concrete is exposed to extraneous moisture and additionally to aggravating factors, such as de-icing salts, freezing and thawing or wetting and drying in a marine environment

The effect of environmental factors on reactions involving carbonate aggregates is less well understood, but probably E1 can be differentiated from E2 and E3.

More detail on the factors affecting the environmental categorisation is given in Table 2.

2.3 The Level of Precaution

The structural and environmental categorisation is combined into the level of precaution in Table 3, where four levels of precaution are identified:

P1. No special precautions against AAR
P2. Normal level of precaution
P3. Special level of precaution
P4. Extraordinary level of precaution

3 Specification to Avoid Damaging Alkali Reactions in Concrete Containing Carbonate Aggregates

The key issue in combating such alkali reactions is identifying potentially susceptible aggregates. In a particular region the existence of such aggregates will generally be known; if not the procedures given below should be followed.

Table 2 Environmental classes

Environmental class	Description	Environment of concrete (see table Notes)
E1	Dry environment protected from extraneous moisture	• Internal concrete within buildings in dry (1) service conditions
E2	Exposed to extraneous moisture	• Internal concrete in buildings where humidity is high; e.g. laundries, tanks, swimming pools • Concrete exposed to moisture from the external atmosphere, to non-aggressive ground or immersed in water. • Internal mass concrete should be included in this category (2)
E3	Exposed to extraneous moisture plus aggravating factors	• Internal or external concrete exposed to deicing salts • Concrete exposed to wetting and drying by seawater (3) or to salt spray • Concrete exposed to freezing and thawing whilst wet • Concrete subjected to prolonged elevated temperatures whilst wet • Concrete roads subject to fluctuating loads

Notes
1. A dry environment corresponds to an ambient average relative humidity condition lower than 75 % (normally only found inside buildings) and no exposure to external moisture sources.
2. A risk of alkali-silica reaction exists for mass concrete elements in a dry environment because the internal concrete may still have a high relative humidity. Vulnerable mass concrete elements are those with a least dimension of 1 m or more.
3. Concrete constantly immersed in seawater does not suffer a higher risk of ASR than a similar element exposed to humid air, buried in the ground, or immersed in pure water, because the alkali concentration of sea water is lower than the alkali concentration of the pore solution of most concretes, and the penetration of chloride ions is usually limited to a few centimetres.

Table 3 Determination of level of precaution

	Environment category (see Table 2)		
	E1	E2	E3
Category of structure (see Table 1)	Level of precaution		
S1	P1	P1	P1
S2	P1	P2	P3
S3	P2	P4	P4

3.1 Identification of Potentially Deleterious Carbonate Rocks

Specific techniques to assess potentially reactive carbonate rocks are given in AAR-0 Annex A. Any carbonate rock about which there is concern is first examined petrographically using the techniques described in AAR-1. If this examination identifies the possibility of alkali-reactivity, the aggregate should be further assessed using the combined tests described in AAR-2 and AAR-5. The possible outcomes are as follows:

(a) AAR-2 expansion exceeds the criteria in AAR-0
AAR-5 expansion is less than AAR-2 by at least 0.01 %
• *There is potential for a deleterious alkali-silica reaction and the precautions in AAR-7.1 should be applied.*
(b) AAR-2 expansion exceeds the criteria in AAR-0
AAR-5 expansion equals or exceeds AAR-2
• *There is potential for a carbonate reaction and the precautions in* Sect. 3.2 *should be applied.*
(c) AAR-2 expansion is less than the criteria in AAR-0
AAR-5 expansion is less than AAR-2 by at least 0.01 %
• *No special precautions against alkali reactions are required.*
(d) AAR-2 expansion is less than the criteria in AAR-0
AAR-5 expansion equals or exceeds AAR-2
• *There is potential for a carbonate reaction and the precautions in* Sect. 3.2 *should be applied.*

3.2 Precautionary Measures in Concrete Containing Potentially Reactive Carbonate Rocks

Although to maintain consistency with AAR 7.1 the four levels of precaution are maintained, in practice the understanding of the reactions in potentially susceptible carbonate rocks and the effectiveness of different precautionary measures does not at present permit distinction in the measures to be taken between precautionary levels P2, P3 or P4.

The options for controlling alkali reactions in such dolomitic carbonate aggregates are much more limited than is the case for alkali-silica reactions. The primary strategy, given the present level of knowledge and understanding is focussed on identifying and avoiding the use of such aggregates. However, experience suggests that some dolomitic carbonate aggregates may be useable in concrete in which the cement contains high replacement levels of blastfurnace slag.

3.2.1 Level of Precaution P1

No special precautions against damaging alkali reactions are necessary.

Ensure that appropriate standards and guidance are followed for the specification of the concrete and good practice employed in its placing and curing.

Note: *If this level of precaution is adopted, some damage from AAR is possible. Therefore the structure must be able to withstand this and the level of damage must be acceptable to the owner.*

3.2.2 Levels of Precaution P2, P3 and P4

At present, only two precautionary measures, either MC1 or MC2, are available for application, with MC1 being a more established option.

Precautionary measure MC1:
Identification and avoidance of use of susceptible aggregates.

This is appropriate for either outcomes 3.1 (b) or 3.1 (d) above. It can be applied on the basis solely of the AAR-2/5 test results or can be confirmed by further testing using either AAR-3 or AAR-4.1.

Precautionary measure MC2:
Use of high replacement levels of blastfurnace slag in the cement.

There is evidence from work in Austria that high replacement levels of blastfurnace slag in the cement can be effective in controlling some reactions involving carbonate aggregates. The effectiveness of this option in the case of a particular aggregate should be assessed using a performance test.

Notes:

1. A suitable performance test is being developed by RILEM.
2. Blastfurnace slag replacement has been found to be ineffective in preventing damage in concrete containing some North American carbonate aggregates [3].

References

1. Gillott, J E, Swenson, E G, 1969, Mechanism of the alkali-carbonate rock reaction. *Quarterly Journal of Engineering Geology*, 7–24.
2. Katayama T, 2004, How to identify carbonate rock reactions in concrete, *Materials Characterization*, (53/2-4), Special Issue (29): 85–104.
3. Thomas, M D A, Innis, F A, 1998, Effect of slag on expansion due to alkali-aggregate reaction in concrete, ACI Materials Journal, 95, 716–724.

RILEM Recommended Specification: AAR-7.3 Preliminary International Specification to Minimise Damage from Alkali-Aggregate Reactions in Concrete—Part 3: Concrete Dams and Other Hydro Structures

1 Introduction

AAR 7.1 provides a model specification for minimising the likelihood of damage in concrete structures that are expected to have a normal service life and where the possibility of premature deterioration, although undesirable, could be managed. There are dams and other hydro structures, however, where the expected service life is very long, probably more than 100 years, and the consequences of premature failure are unacceptable.

In such structures, especially stringent precautions will be needed and some mechanisms of deterioration may come into play that are not present in normal structures and which are not yet fully understood. This RILEM Recommendation gives the most up to date advice on avoiding damage from alkali-aggregate reactions in such structures, but it must be accepted that this is a field where some uncertainties remain. The ideas underpinning this Recommendation have initially been put forward in papers presented at the 14th ICAAR (2012) [1] and at Hydro 2012 in Bilbao [2].

2 General Principles for Avoiding Damage from Alkali Reactions

AAR 7.1, the International Specification to minimise damage from alkali reactions in normal concrete structures, has the basis that alkali-silica reactions take place when certain susceptible forms of silica in the aggregates react with the alkaline pore solution in the concrete. This produces an alkali-silicate gel which absorbs water producing an expansive force which can crack and expand the concrete at the micro-level and cause deformations and structural cracks at the macro-level. The concrete should, therefore, be designed to avoid the simultaneous presence of:

P.J. Nixon and I. Sims (eds.), *RILEM Recommendations for the Prevention of Damage by Alkali-Aggregate Reactions in New Concrete Structures*, RILEM State-of-the-Art Reports 17, DOI 10.1007/978-94-017-7252-5_10

- A sufficiently alkaline pore solution in the concrete
- A critical amount of reactive silica
- A sufficient supply of water

To achieve this with the minimum financial and environmental cost, the precautions should be tailored to the nature and service life of the structure. Therefore, the development of the precautions should take the following form:

1. Determining the necessary level of precaution;
2. Undertaking recommendations according to the level of precaution required.

2.1 Necessary Level of Precaution

This will depend on the combination of the structural needs and planned service life of the structure and the environment to which the concrete is exposed. The structural needs and service life are characterised in terms of the adverse consequences associated with any damage, whilst the environment is determined, above all, by the supply of moisture.

In smaller structures, a supply of moisture extraneous to the concrete itself is necessary, but in larger structures sufficient moisture is probably retained after hydration to initiate, and in many cases maintain, reaction and facilitate on-going expansion. Other, aggravating, factors that will influence the severity of AAR-induced damage include the application of sodium chloride based de-icing salts, exposure to seawater and the synergistic effects of freezing and thawing or other deteriorative mechanisms. From a consideration of these factors AAR-7.1 defines four levels of precaution, P1 to P4, with P4 being the most demanding.

2.2 The Available Precautionary Measures

AAR 7.1 describes four types of precautionary measures which may be applied, designated M1, M2, M3 and M4. Of these, the precautionary measures M1 and M2 are best established and are recommended by RILEM. There is a detailed discussion of how to apply these measures in AAR 7.1.

M1: Measures to restrict the alkalinity of the pore solution.

There are several ways of achieving this:

- Limiting the alkali content of the concrete
- Use of a low-alkali cement
- Inclusion in the concrete of a sufficient proportion of a low lime-fly ash, other pozzolana demonstrated to be effective, or ground granulated blastfurnace slag

M2: Measures to avoid the use of a reactive aggregate combination
(i.e. total aggregate combination of coarse and fine aggregates).

In the context of ASR, reactive silica occurs almost exclusively in the aggregate. Therefore to make use of this precautionary measure, the RILEM Recommended test methods should be used, to identify and thereby avoid, if possible, "reactive aggregate" combinations. It is important to test different aggregate proportions as with some aggregates small amounts of reactive silica can be more damaging (the "pessimum" effect). In many cases of dam construction, particularly in remote locations, the available choices of aggregate are limited, such that this may not be a practical or economic option.

M3: Measures to reduce the access of moisture and maintain the concrete in a sufficiently dry state.

In water retaining structures, such as dams, reducing the access of water may be difficult or impracticable. Nevertheless there may be circumstances where additional protection can be gained by such measures and this is discussed in Sect. 5.

M4: Measures to modify the properties of any gel such that it is non-expansive.

This, which at present is essentially limited to the use of lithium salts, is not yet well proven for long-term effectiveness and should be used with caution and only after trials to establish the effective dose for the concrete mix in question.

3 Application to Large-Scale, Long Service Life Structures Such as Dams

3.1 Necessary Level of Precaution

In the case of concrete dams, their structural importance, extremely long planned service lives and the fact that they are permanently exposed to moisture inevitably lead to the conclusion that the highest levels of precaution are needed. In AAR 7.1, this is termed P4 and the recommended measures to avoid ASR damage are:

Level of precaution P4: *This extraordinary level of precaution is only needed in structures where the consequences of any deterioration are unacceptable. In general it will necessitate the combined application of at least two of the four precautionary measures (M1, M2, M3 or M4), although it will be shown that this choice is often more limited for dams and similar structures. Additionally the concrete should be designed to resist any aggravating factors such as freezing and thawing whilst wet, de-icing salts or wetting and drying in a marine atmosphere.*

3.2 Precautionary Measures: Special Factors Affecting Large-Scale, Long-Service Structures

When taking measures to meet the requirement P4 in large-scale, long service structures, there are several special factors that need to be taken into account:

- slow reactions, which might cause limited harm and/or terminate reasonably quickly within smaller concrete structures, may continue indefinitely in very large structures, sometimes gradually leading to deformations and deterioration that threaten the integrity and serviceability of the structure. Conventional (invariably accelerated) test methods for aggregate reactivity may not identify such reactions;
- unlike small concrete specimens in accelerated laboratory tests, the scale of very large structures minimises the scope for the lixiviation (leaching) of alkalis;
- the need for structural safety is paramount in such structures;
- by definition the concrete will be wet (although some measures might reduce water ingress, much of the concrete will never dry and, in practice, it is safer to assume it will always be wet);
- dams are often built in relatively remote areas where the choice of materials is limited and, in particular, there will be strong economic and environmental reasons for using local aggregates;
- the horizontal restraint imposed by the contact of the dam with the solid rock at its extremities can lead to bending or bowing of the dam if the concrete expands, especially in arch dams or with gravity dams that have a curved axis;
- any concrete expansion can lead to the serviceability of a dam being threatened by distortion of connections to spillways, gates, machineries and other installations.

Therefore the advice given in AAR-7.1 for this situation (P4) needs to be modified in several ways.

Firstly, as mentioned above, the facts of construction in remote areas may restrict the choice of materials, especially aggregates, which could lead to difficulties in applying measure M2.

Secondly, the continuously wet environment and extreme longevity may lead to special problems. In particular, there is concern about the possibility of some types of aggregates acting as a reservoir for alkalis, which then enable the reaction to be maintained over a longer time scale than would be expected in structures where the alkalis come mainly from the cement. Even if the aggregate itself does not contribute alkalis, the large volume of concrete will provide a reservoir of alkalis which may be concentrated to damaging amounts by moisture migrations within the concrete. Additionally, there is also the possibility of alkali recycling, as earlier-formed alkali-silica gel gradually alters and can re-release its alkalis [3].

Overall, the preference for very long term protection is still for the application of at least two separate precautionary measures. However, if this is impractical then the more rigorous application of one precautionary measure, combined with planning of the design and construction to minimise the effects of any unavoidable expansions, will be required to give a high level of protection.

4 Prevention by Choice of Materials

As explained above, the usual precautionary measures are designed to avoid either or both of a high alkalinity in the concrete (M1) and the presence of a critical amount of reactive silica in the aggregate (M2). In respect of these measures, the combination of non-reactive aggregates with a reasonably low alkalinity in the concrete appears to be an ideal solution if aggregate choices are available.

The initially available alkalinity in the concrete mainly arises from the alkalis in cements and is modified by the inclusion of slags and pozzolanic materials (but see discussion below of the subsequent release of alkalis from aggregates and/or alkali recycling). As cements, slags or pozzolanic materials will inevitably have to be imported to the construction site and are, in any case, a less bulky component of the concrete than the aggregate, their choice will depend on what is available in the region. However, careful choice of the cement and use of mineral additions or 'supplementary cementitious materials' (SCMs), such as slags and pozzolanas, can give very powerful protection to the concrete. In some regions, such as Brazil, the use of large proportions of SCM is becoming universal for the concrete in dams, but effectiveness varies and some SCM varieties may themselves release alkalis in the long-term. Nevertheless, although current indications are certainly encouraging, it remains to be established conclusively that SCMs will always limit long-term reaction prospects in very large structures.

In the case of aggregates for dams, it is usually necessary, for economic and environmental reasons, to use local materials. Often this will mean opening new quarries, so there may not be a history of use and effective testing becomes particularly important. It is therefore important that an assessment of the aggregate should be carried out at the feasibility stage and the long planning periods involved in large dam projects should be utilised to allow the testing programme to be planned and carried out effectively, including the use of long-term tests. This contrasts with many construction projects, where there is not enough time for adequate assessment.

4.1 Cementitious Materials

As summarised above, the alkalinity of the pore solution in the concrete can be kept to a level where damaging reaction with silica in aggregates is unlikely by limiting the alkali content of the concrete mix, or by inclusion in the concrete of a sufficient proportion of a low-lime fly ash, or other pozzolana demonstrated to be effective, or ground granulated blastfurnace slag.

AAR 7.1 advocates that the limit on the alkali content of the concrete and the minimum amounts of ash or slag used should be based on the reactivity of the aggregate. In the case of these large-scale long-service structures, extra protection for the concrete can be obtained by ensuring that the alkali level in the concrete is kept to a particularly low level, regardless of the aggregate reactivity and by additionally using high levels of a low-lime fly ash or good quality ground granulated slag.

If it is assumed that the aggregate is high reactivity, according to AAR 7.1, the limits and quantities shown in Table 1 would apply.

Meeting these low alkali levels in the concrete may not be too onerous as the cement content in large-scale structures is likely to be comparatively low, to avoid heat rise problems, and similarly ashes and slags are often used in such structures for the same reason.

If it is not possible to use the two precautionary measures M1 and M2, then emphasis should be given to maximising the effectiveness of precautionary measure M1 through the combined use of a concrete with a low-alkali content and sufficient ash or slag to meet the limits shown in Table 1.

It is noteworthy that there are currently no cases known to the RILEM TC of damage to dams by alkali-aggregate reactions when the concrete has contained these levels of good quality fly ash or slag. If such a combination of low alkali level in the concrete and sufficient ash or slag are used, it will be important to ensure that the basic strength gain of the concrete is sufficient, as the reactivity of the ash or slag is aided by a higher level of alkalis in the cement.

Other pozzolanic materials, such as natural pozzolanas, silica fume or metakaolin, can give good protection, but their use is less well established and cannot yet be recommended in critical structures like dams unless their effectiveness has been demonstrated by a performance test, which will also establish the optimum proportion required. It is likely that the ultrafine materials (silica fume or metakaolin), as well as being pozzolanic, might be especially effective in densifying concrete and thereby helping to inhibit alkali release, recycling and movement, although these potential benefits need to be verified by further research. Cases have been reported in which some silica fume used at proportions of less than about 10 % by mass of cement has only a delaying effect on AAR, rather than an enduring preventative

Table 1 Limits for concrete containing high reactivity aggregate

Alkali level in concrete[a] (Na_2O eq.)	≤2.5–3.0 kg/m^3
Low-lime fly ash (<8 % CaO and <5 % Na_2O eq.)	>40 %[b]
Ground granulated blastfurnace slag (<1.5 % Na_2O eq.)	>50 %[b,c]

[a]Total releasable alkalis from all mix constituents, including cement, any SCMs, any admixtures, mix water and all the aggregates

[b]% by mass of total cementitious material. Provided that these minimum proportions are used, and subject to local experience with particular materials, the alkali content of the fly ash or slag need not be included in the calculation of the "reactive" alkalis in the concrete

[c]In France, a minimum proportion of 60 % ggbs is being applied

influence [4, 5]. There is also growing evidence that tripartite blends (cement plus two types of addition or SCM) can be particularly effective [6].

4.2 Aggregate Assessment

The suite of RILEM test methods together with overall guidance in AAR-0 allows a reliable assessment of the reactivity of aggregates. In summary, the assessment should start with a petrographic examination according to AAR-1.1. This will enable the choice of the best test methods for a full assessment and will also reveal the uniformity of the aggregate sources and determine how the sampling for subsequent tests should be undertaken. Then, optionally, AAR-2, the accelerated mortar-bar test, can be used for initial screening, leading on to the longer-term concrete prism tests in AAR-3.1 (38 °C storage) and AAR-4.1 (60 °C storage). If there are carbonate aggregates, these can be assessed using the procedure in AAR-0 and the combined test methods of AAR-2 and AAR-5.

The inter-laboratory tests carried out by RILEM [7, 8] and the PARTNER project [9], which assessed the RILEM methods for use in Europe, both showed that the AAR-3.1 and AAR-4.1 methods can reliably identify non-reactive aggregate combinations and aggregate combinations that will become reactive in normal timescales. The methods were least reliable in identifying aggregate combinations that were marginally reactive or reacted over long timescales and in the case of long-service structures this could be important. Of the two, AAR-4.1, the accelerated (60 °C) concrete test, was more effective in identifying these marginal/slow reactions in the normal test period, although the effectiveness of AAR-3.1 for dams could be improved by extending its normal 1 year test period to 2 years. When the test period was extended in this way, the AAR-3.1 method was found by the PARTNER project to show the best correspondence with field experience. In the case of important, long service structures there is a good argument for carrying out both methods.

4.3 Expansion Limits: Special Factors When Testing for Dams

In structures like dams, there is a need to reconsider the normal criteria used to interpret expansion test results. These normal criteria are currently that for AAR-3.1, expansions exceeding 0.05 % at 1 year indicate the possibility of a harmful alkali-aggregate reaction, whilst in AAR-4.1 the limit is 0.03 % at 15 weeks.

The evidence from the comparison of the expansions in laboratory tests with those of long-term field trials of large concrete specimens shows that, when there is an alkali-aggregate reaction, the eventual expansions in the large specimens are much greater than those in the laboratory, probably because of alkali leaching from the smaller laboratory specimens. Consequently, when establishing an expansion test limit, a safety margin should be provided to cover this discrepancy. Moreover, some mixes that do not expand in the laboratory tests may eventually expand in the field [10].

However, it is essential to recognise that even low levels of expansion can lead to significant effects in a dam: 0.05 % expansion (the normally applied expansion limit in AAR-3) over 100 m of dam concrete represents 50 mm of movement, probably more than can be tolerated. Thin arch dams are especially sensitive to expansion, whereas some other designs of dams may be more tolerant [11].

There is, therefore, a strong argument for reducing the expansion test criteria to values that are consistent with the tolerable expansion in the prototype structure, although this may be below the precision level of the accelerated methods. The tolerable expansion limit in a dam will vary, depending on the dimensions, configuration and related equipment. In cases where the use of reactive aggregates is unavoidable, an assessment of the tolerable expansion limit over a realistic service life of the structure should be made and compared with the test criteria, recognising the limits of precision (precision data are provided in AAR-0). If the tolerable expansion criteria cannot be satisfied by testing, then it will be necessary for the designers to modify the design of the concrete structure so that the potential long-term expansion can be tolerated.

As one of the major concerns is the detection of long-term, slow reactions, there is also reason to extend the test periods; to 2 years, if possible, in the case of AAR-3.1 and to 20 weeks or preferably longer for AAR-4.1. In such cases, where a tolerable expansion limit is beyond testing precision or not known, then it is also necessary to examine the expansion curves for any signs of continuing expansion, because the expansion might not be complete in the test specimen even after these longer test periods.

The currently suggested maximum expansion limits for concrete for very large, long-service structures are therefore:

AAR-3.1: 0.03 % at 1 year, and/or 0.04 % at 2 years

AAR-4.1: 0.02 % at 15 weeks, and/or 0.03 % at 20 weeks or longer (e.g. in the French concrete performance test, where the 0.03 % limit is applied at 1 year [12]).

These suggested test limits represent the current state of the art for providing a robust assurance for prevention of adverse expansion effects in very large, long-service, concrete dams and similar hydraulic structures.

Overall, an expert review of the combined results from petrographic examinations and the results of AAR-3.1 and AAR-4.1 tests, considering the potential impacts of expansions of at least the test limits on a specific prototype structure, recognising its dimensions, geometry and required service life, will be required to give the most reliable assessment.

4.4 Performance Tests

The ideal would be to assess the behaviour of the actual cement/aggregate combination in dependable tests linked to realistic performance limits. At present, the RILEM TC is developing a performance test based on the AAR-3 (38 °C) concrete prism method. Meanwhile, in France, the 60 °C concrete prism method (similar to AAR-4.1) is being applied in practice as a performance test [12].

In the case of mixes containing pozzolanic materials, it is particularly important that such performance tests are backed up with long-term outdoor field tests (with large concrete blocks of at least 1 m dimensions) or by site experience, because some accelerated laboratory tests, using elevated humidity and temperatures, may give unrealistically optimistic results for the effectiveness of pozzolanic materials in combating ASR expansion (the test conditions can enhance the activity of the SCMs). Similarly, it is not yet known how effective such a performance test would be in identifying the slow and later reactions that are of concern in these very large structures. It could be argued that, in such cases, a higher temperature applied during the performance test, such as that used in the AAR-4.1 (60 °C) method, might be useful in accelerating otherwise very slow reactions.

Another approach, which has been applied in France in cases when non-reactive aggregates are not available, is to use a performance test to determine the 'alkali threshold' for the particular aggregate/cement combination (using aggregates representative of the ones to be used in the structure) and then apply a safety factor to the alkali level actually used in the structure. The alkali threshold is the lowest alkali level in the concrete at which a damaging expansion is found in tests and a method for determining this is given in AAR 3.2. In France the performance test on concrete is done at 60 °C, according to AFNOR NF P 18-454 [13]; this is the method on which AAR 4.1 is based, but at present RILEM is recommending the AAR 3.2, 38 °C, method for this purpose. Once the alkali threshold has been determined, the project mix can be designed with a safety factor in the form of a lower alkali level. Depending on the criticality of the structure and the confidence with which the alkali threshold has been determined, a safety factor can be applied by reducing the alkali level in the project mix by between 1.0 and 2.0 kg/m^3 Na_2Oeq. below the alkali threshold. This seems a promising approach and it will also give some protection against possible long-term alkali contribution from aggregates.

A further idea, which is being suggested in North America, is the establishment of large project-specific field site specimens made with materials and mitigation measures that are expected to be used and as far in advance of the dam construction as possible [14, 15]. There is good evidence that in the long-term such large specimens will give a more reliable guide to the behaviour of real structures than laboratory specimens. However, given that such tests are at ambient temperatures, rather than being 'accelerated' by elevated temperatures, even if a lead time of

5 years could be managed in many cases, some means of assessment over and above expansion measurements will be needed to give early warning of any developing reactions. It is possible that detailed microscopical examination, using optical or electron microscopy, of cores taken from the specimens could help with this (microscopic evidence of ASR becoming discernible before any measurable deformation or visible damage to the structure), but some form of validation using existing specimens will be needed to demonstrate the effectiveness of such a method.

4.5 Alkalis in Aggregates

Many minerals in aggregates, for example feldspars, can contain significant amounts of alkalis. Although most of these alkalis are securely chemically bound within stable minerals and thus not releasable, given that the aggregate makes up such a high proportion, perhaps 75 % by weight, of the concrete, even a small proportion of these alkalis have the potential to contribute significant amounts of alkali to the pore solution.

In some cases, geological alteration or weathering of minerals, such as feldspar, can greatly enhance potential releasability of alkalis. Generally, however, even these potentially releasable alkalis are not readily soluble in the pore solution, at least in timescales appropriate to most concrete structures. In the case of structures designed for very long service lives, however, there is concern that the aggregates can contribute alkalis in the long-term and negate the precautions taken in designing the concrete.

Unfortunately, whilst it is now certain that some aggregates can release alkalis in this way, there is presently no consensus on how to test for releasable alkalis in aggregates. The result of testing for these alkalis depends very heavily on the ways in which the aggregate samples are prepared for testing, the extraction solution that is used and above all the duration of the extraction period. Various solutions have been used experimentally, including water, alkaline, acid, saturated lime and so on, and the results vary widely. Moreover, it has proved difficult to correlate the results of laboratory tests to what happens in field concrete. RILEM is developing a standardised test, AAR-8, to determine the proportion of alkalis that might be released from an aggregate in concrete, probably employing an extraction solution similar to that of concrete pore solution. A first draft of AAR-8 should become available within the next year or so, with practical research to establish criteria for use with AAR-8 being completed during the period of the current RILEM TC AAA.

In the meantime, the recommended approach is to use a petrographic examination to identify aggregate mineral compositions that are believed to be potentially susceptible to alkali release and, if possible, to avoid the use of aggregates containing significant amounts of such minerals. Most concern has focused on feldspars, which are common rock-forming minerals, especially when geological alteration or weathering has initiated degradation of the feldspar, with associated formation of secondary clay minerals. It is not practicable or appropriate routinely to avoid using aggregates containing feldspar, as it is an abundant rock-forming mineral, but such aggregates should be avoided, if possible, when there is a significant degree of associated alteration or weathering. Many sedimentary rocks will also contain some clay minerals, which could be a source of releasable alkalis. Experience has shown that many metamorphic rocks (such as gneiss and mica schist) may also release alkalis.

4.6 *Synergistic Effects*

There is evidence that other mechanisms that produce expansive forces in concrete, such as freezing and thawing, internal sulfate attack (ISA) and delayed ettringite formation (DEF), can have a synergistic effect with alkali-aggregate reactions; each making the other worse [16, 17]. Avoidance of these effects, important in itself, should therefore be seen as an integral part of avoiding long-term alkali-aggregate damage in very long-service structures. Thus, strenuous endeavours to avoid AAR, whilst obviously laudable, should not cause all of the other potential threats to concrete durability to be overlooked or considered to be of secondary importance.

5 Prevention by Structural Design

From a structural point of view, the designer of a major structure always prefers the choice of a non-reactive concrete formula. It is only when the concrete unavoidably carries some AAR risks and/or when the risk of long-term AAR is difficult to assess that the designer may need to introduce special aspects of structural design to mitigate the potential consequences of any expansion.

The options for reduction of long-term moisture content in the concrete are limited in large dams and similar hydraulic structures. Although in most cases sufficient residual moisture will be present to feed the reaction and expansion, it would clearly be prudent to provide some weather protection and drainage. To this end, the structure should be designed so as to avoid, to the greatest extent possible, water accumulation and stagnation zones as well as internal flow paths by providing slope profiles and shapes that allow for rapid water run-off. In the case of dams, the upstream face is generally in permanent contact with water, but the internal drainage systems provided for uplift control can assist in minimising water accumulation inside the mass of concrete and in channelling seepage water through shafts and galleries.

In some cases, such as relatively thin dam and spillway structures, it could be useful to apply a membrane or a coating on the exposed faces of the structure, to limit water penetration into the concrete. The coating could be paint, a thin protective layer or a thick layer. However, this type of coating only usually maintains its efficiency over a limited life cycle, so that a number of coating replacements will be needed during the service life of the structure. The effects of impervious surface coatings on the internal moisture condition and behaviour of the concrete is also uncertain. In some cases, the coatings may limit evaporation and surface drying. In some smaller structures, repair methods using impervious coatings have sometimes made matters worse. However, in some circumstances, including cases where reservoirs can be emptied and waterproofing applied to the upstream face of as concrete dam only after a protracted period of drying, continued expansion caused by AAR is reported to have been significantly reduced [18].

Another pragmatic way to mitigate adverse AAR effects is to make allowance for possible expansion. For example, it is possible to design sufficiently wide joints between blocks of concrete to accommodate the expansion of the concrete. In some cases, it could be beneficial to use 3D detailed reinforcement or prestressing bars and tendons in the construction of some critical elements that have high levels of tension stresses, to control cracking.

Additionally, during the service life of the structure, it is necessary systematically to inspect the parts deemed to be critical, so as to detect any cracks that may appear and allow water to penetrate into the concrete.

Comprehensive instrumentation from the outset, coupled with structural modelling, is also essential, to quantify deformations and identify the effects of potential deformations and provide warning of the need for remedial interventions.

6 Summary of Precautions for Large-Scale, Long-Service Structures

These structures are potentially vulnerable to damaging alkali-aggregate reactions because of their longevity and environment. In terms of the RILEM International Specification, AAR 7.1, the necessary level of precaution is P4, the highest category, and classified by AAR 7.1 as an Extraordinary level of Precaution.

Moreover, it may not be straightforward to apply the two separate precautionary measures that are recommended for this level of Precaution, because of the frequent need to use locally available materials, particularly the aggregates, which may restrict the available choices. However, although the application of at least two separate precautionary measures is still the preferred solution, where necessary, a high level of protection can be achieved by the more rigorous application of one precautionary measure, combined with design and construction measures to minimise the effects of any expansion.

These more rigorous precautionary measures are summarised in Table 2.

Table 2 Summarised precautionary measures for massive long-service structures

CM1	Use of non-reactive aggregate combination	Expansion: (AAR-3.1) <0.03 % at 1 year, and/or <0.04 % at 2 years (AAR-4.1) <0.02 % at 15 weeks, and/or <0.03 % at 20 weeks or longer—see earlier
CM2	Avoidance of releasable alkalis in aggregates	On basis of petrographic examination until AAR-8 and its criteria are available
CM3	Alkali level in concrete	<2.5 kg/m^3 (Na_2O eq.)[a] or
		Safety factor of 1, 1.5 or 2.0 kg/m^3 (Na_2O eq.) below the alkali threshold determined by a performance test on the actual concrete composition
CM4	Use of low-lime and low-alkali fly ash (<8 % CaO and <5 % Na_2O eq.) or	At least 40 %[b]
	Use of ground granulated blastfurnace slag (<1.5 % Na_2O eq.)	At least 50 %[b]
CM5[c]	Design and construction measures	Drainage, protection by membranes or coatings, detailing of reinforcement or prestressing of critical elements, inclusion of expansion joints, monitoring and indicative field specimens

[a]Including releasable alkalis from all concrete constituents
[b]By mass of cementitious material
[c]CM5 only to accompany and complement the other measures

References

1. Sims, I, Nixon, P J, Godart, B, 2012, Eliminating alkali-aggregate reaction from long-service structures, In: Folliard, K J (ed) *Proceedings of the 14th International Conference on Alkali-Aggregate Reaction*, 20–25 May 2012, Austin, Texas, USA.
2. Sims, I, Nixon, P J, Godart, B, Charlwood, R, 2012, Recent developments in the management of chemical expansion of concrete in dams and hydro projects - Part 2: RILEM proposals for prevention of AAR in new dams, In: *Proceedings, Hydro 2012 International Conference, Innovative Approaches to Global Challenges*, 29–31 October 2012, Bilbao, Spain, available as a CD.
3. French, W J, 2005, Presidential address 2003: Why concrete cracks - geological factors in concrete failure, Proceedings of the Geologists' Association, 116(2), 89–105.
4. Duchesne, J, Bérubé, M-A, 2000, Long- term effectiveness of supplementary cementing materials against ASR, In: Bérubé, M-A, Fournier, B, Durand, B (eds) Proceedings, 11th International Conference on Alkali-Aggregate Reaction, Québec City, Canada, 613–622.
5. Gudmundsson, G, Halfdanarson, J, Möller, J, 2000, Effect of silica fume properties on mitigation of ASR reactivity in concrete, In: Bérubé, M-A, Fournier, B, Durand, B (eds) Proceedings, 11th International Conference on Alkali-Aggregate Reaction, Québec City, Canada, 643–651.
6. Bleszynski, R F, Thomas, M D A, Hooton, R D, 2000, The efficacy of ternary cementitious systems for controlling expansion due to alkali-silica reaction in concrete, In: Bérubé, M-A,

Fournier, B, Durand, B (eds) Proceedings, 11th International Conference on Alkali-Aggregate Reaction, Québec City, Canada, 583–592.
7. Nixon, P J, Sims, I, 2000, Universally accepted testing procedures for AAR: the progress of RILEM Technical Committee 106, In: Bérubé, M-A, Fournier, B, Durand, B (eds), Proceedings, 11th International Conference, Alkali-Aggregate Reaction in Concrete, Québec City, Canada, 435–444.
8. Sims, I, Nixon, P J, 2006, Assessment of aggregates for alkali-aggregate reactivity potential; RILEM International Recommendations, In: Fournier, B (ed), Marc-André Bérubé Symposium 8th CANMET International Conference on Recent Advances in Concrete Technology).
9. Lindgard J, Nixon P J, Borchers I, Schoenborg B, Wigum J B, Haugen M, Akesson U, 2010, The EU PARTNER project – European standard tests to prevent alkali reactions in aggregates: Final results and recommendations, Cement and Concrete Research, 40, 611–635.
10. Thomas M, Dunster A, Nixon P. J, Blackwell B, 2011, Effect of flyash on the expansion of concrete due to alkali-silica reaction – exposure site studies, Cement and Concrete Composites, 33, 359–367.
11. Amberg F., Performance of dams affected by expanding concrete, International Symposium on Dams and Reservoirs under Changing Challenges – 79 Annual Meeting of ICOLD, Lucerne, Switzerland, 1 June 2011.
12. Bollotte, B, 1992, Development of an accelerated performance test on concrete for evaluation of its resistance to AAR, In: 9th International Conference on Alkali-Aggregate Reaction in Concrete, London, Conference Papers, Volume 1, 110–116.
13. AFNOR, 2004, Reactivity of a concrete formula with regard to the alkali-aggregate reaction, Performance test, AFNOR NF P 18–454, plus AFNOR FD P 18-456 (Interpretation criteria for NF P 18-454), Association Française de Normalisation, Paris, France.
14. Rogers, C, 2012, Personal Communication.
15. Thomas, M, 2012, Personal Communication.
16. LCPC, 2009, Recommendations for preventing disorders due to Delayed Ettringite Formation.
17. Martin, R-P, Bazin, C, Toutlemonde, F, 2012, Alkali aggregate reaction and delayed ettringite formation: Common features and differences, In: Folliard, K J (ed), Proceedings, 14th International Conference on Alkali-Aggregate Reaction, Austin, Texas, USA.
18. Matos, S, Ferreira, I, Camelo, A, Gomes, J P, Scuero, A, Vaschetti, G, Vale, J M, 2013, Pracana Dam 20 years after rehabilitation - observed behaviour of the dam and of its exposed geomembrane, ICOLD 2013 International Symposium, Seattle, USA.

MIX
Papier aus verantwortungsvollen Quellen
Paper from responsible sources
FSC® C105338
FSC
www.fsc.org

If you have any concerns about our products,
you can contact us on
ProductSafety@springernature.com

In case Publisher is established outside the EU,
the EU authorized representative is:
Springer Nature Customer Service Center GmbH
Europaplatz 3, 69115 Heidelberg, Germany

Printed by Libri Plureos GmbH
in Hamburg, Germany